James Ross

On Aphasia

Being a Contribution to the Subject of the Dissolution of Speech From

Cerebral Disease

James Ross

On Aphasia
Being a Contribution to the Subject of the Dissolution of Speech From Cerebral Disease

ISBN/EAN: 9783744759007

Printed in Europe, USA, Canada, Australia, Japan

Cover: Foto ©berggeist007 / pixelio.de

More available books at **www.hansebooks.com**

ON APHASIA.

ON APHASIA:

BEING A

CONTRIBUTION TO THE SUBJECT OF THE DISSOLUTION

OF SPEECH FROM CEREBRAL DISEASE.

BY

JAMES ROSS, M.D., LL.D. ABERD.,

Fellow of the Royal College of Physicians of London, and Senior Assistant Physician to the Manchester Royal Infirmary.

LONDON:

J. & A. CHURCHILL,

11, NEW BURLINGTON STREET.

1887.

PREFACE.

IN the following pages is collected, with a few alterations and additions, a series of papers that appeared during the past year in the *Medical Chronicle*. My main object in writing these papers was to record several interesting and instructive cases of aphasia which had come under my own observation, but inasmuch as in none of my cases did the symptoms observed come to the test of a *post-mortem* examination, I could not enter profitably, either on an analysis of the clinical phenomena, or on an interpretation of them in terms of structure, without availing myself largely of the labours of others. Although previous observations on aphasia will be found described here with tolerable fulness, yet, at the same time, the work does not pretend to be an exhaustive monograph, nor is the subject treated in a strictly systematic manner. In dealing with the opinions of others, I have endeavoured to separate what I conceive to be the true portions of discordant theories from the errors with which they are intermingled, and have then striven to combine the true into one harmonious doctrine. That I have succeeded in separating the true from all admixture of error and in producing an unimpeachable theory of aphasia, I will not for a moment maintain; but it would be mere false modesty were I to disclaim all credit for the doctrine set forth in this work, or to deny that I believe it to be more harmonious in all its parts than any which has hitherto appeared. The authors whose works I have laid under contribution are referred to in the body of the work, but I cannot pass over without special mention the names of Broca, Trousseau, and Charcot in France; Kussmaul and Wernicke in Germany; Hughlings-Jackson, Gairdner, Bastian, Broadbent, and Ferrier in this country. These are the great masters of the subject, and any one who undertakes to write upon it ought to make a close study of their works.

It will be observed that the closing part of this volume descends almost to the level of a polemic against Dr. Broadbent's views. The distinguished author will, I hope, believe me when I say that this arises, not because I attach little, but because I attach much, importance to his opinions. Some portions of Dr. Broadbent's work on Aphasia must be simply accepted and not criticised, but other portions are open to question. He belongs to a school of psychologists which manifests a tendency to break up the human mind into numerous faculties, with their corresponding cortical centres, and it is as the highest exponent of this school that his views are here singled out for adverse comment.

JAMES ROSS.

MANCHESTER, JANUARY, 1887.

CONTENTS.

APHASIA.

APHASIA (*à* priv., and *φάσις* speech) means literally loss of the faculty of intelligent speech, or speechlessness. Before defining more accurately the sense in which the term is to be used in the following pages, it is desirable to make a few general remarks on language, and on the nervous mechanisms which the full use of language presupposes. Language, taken in its widest sense, consists of the various means by which animals indicate and appreciate the mental states of one another. Mental states are divided by psychologists into feelings, cognitions, and volitions. Language is frequently expressive of volitions, but inasmuch as volitions are determined by motives, or in other words by the feelings and cognitions, language of the volitions merges itself into that of the other two mental states. Language may, therefore, be divided into that of the feelings, or *emotional* language, and that of the cognitions, or *intellectual* language, and it is to the latter alone that the term speech is truly applicable. The language of the emotions is, as Mr. Darwin has shown, fairly well developed in animals so low down in the scale of being as insects, and it is highly developed in the higher vertebrata. Intellectual language, however, is almost exclusively limited to man. Faint indications of this form of language may undoubtedly be observed in the lower animals, but in them it appears only in a rudimentary form, and seldom, if ever, completely differentiated from emotional language. A rook, for instance, perched on a high tree, utters in quick succession, "Caw, caw, caw," and immediately the neighbouring community of rooks take to flight. It is difficult to deny an intellectual quality to this act, inasmuch as the given signal had indicated the approach of danger to the community of rooks, and the community had adopted the best means, namely, flight, for avoiding the approaching danger. Scanned more closely, however, it will be seen that the signal cry of the rook belongs chiefly to emotional, and only in a very slight degree to intellectual, language. The cry had indicated to the community that the rook giving the signal was the subject of the emotion of fear; the same emotion was communicated to the community at large, and all the rooks immediately took to their inherited mode of escape from enemies. The actions of the rooks were, therefore, determined almost exclusively by emotion, and only in a very subordinate degree by the intellect, or, as Professor Huxley aptly expresses it, by a train of feeling, and not by a train of thought. It is in man alone, therefore, that the language of the intellect becomes more or less completely differentiated from the language of the emotions, and even in him the division between them is by no means always clear and trenchant. When, for instance, a public speaker delivers an oration, only a small part of what he utters can be regarded as speech. The variations of tone, the melodious voice, the graces of attitude and gesture, the charm of elegant and rhythmical language, and the thousand other ways by which a great orator knows how to sway and influence his audience, belong to emotional and not to intellectual language. And similarly with regard to written language. The separate propositions on a printed page belong to intellectual language, but such methods as accent, italics, and notes of exclamation, by which inflection, and emphasis, and wonder are expressed, as well as the rhythm of metrical compositions, and the diction and imagery of poetry, all belong to emotional language. The distinction, however, between emotional and intellectual language is a radical one, and it will be hereafter seen that the former faculty, which is much the more deeply organised of the two, may be almost entirely unaffected while the latter is almost quite lost by disease.

Let us now look more narrowly at the structure of intellectual language. This kind of language gives expression to a thought or cognition, and a thought when expressed has the value even when it does not assume the form of a proposition. If I utter the word "book," it may or may not convey any meaning to the listener. But if any thought is conveyed by the word the listener must understand my intention to be to express that the object I am pointing at, holding in my hand, or thinking about, belongs to the class of things known to him and to me by the name of a book, and the word has to both of us the value of a distinct proposition. A proposition, therefore, is the unit of intellectual language. Language is the instrument of the social state, and in order that it may be a means of intercommunication between animals it must possess to each a *subjective* and an *objective* value, or in other words it must fulfil an *impressive* and an *expressive* function. The signal cry of the rook, already used as an illustration, indicates that the rook which utters it is in a certain mental condition, but this indication would be of no value to the community unless a similar mental condition were excited in all the rooks that listened to the cry. In order that the active rook may be able to communicate his own emotion or thought to the other rooks he must be possessed of outgoing or centrifugal mechanisms, by means of which he is able to give a certain signal ; and in order that the signal may fulfil the function of a language the passive rooks must be possessed of ingoing or centripetal mechanisms, by means of which they are able to appreciate the meaning of the signal. And each rook, in order to be an effective member of the community, must be possessed of both the outgoing and ingoing mechanisms, so that he may, as occasion requires, be able on the one hand to give warning of approaching danger to the other rooks, and on the other to appreciate the meaning of the warning cry when hearing it from another rook. Now, as the outgoing or expressive function always results in a muscular action of some kind, it may be briefly named the *motor* function of language ; and as the ingoing or impressive function presupposes the activity of the ear, eye, or one of the other senses, it may be named the *sensory* function of language. The first step of our analysis therefore shows us that language consists of motor and sensory functions, and that these are regulated by corresponding motor and sensory nervous mechanisms. If the motor function of language, or rather of speech, with which we are now particularly concerned, be further analysed, it will be found to consist of certain articulate sounds and vocalisations constituting articulate or *spoken* speech, certain movements of the hand resulting in *written* speech, and particular movements of the limbs and body constituting *pantomimic* speech. Each of these kinds of speech is regulated by a mechanism in the spinal cord, or medulla oblongata, with its attached nerves, and by a cerebral cortical centre with its attached centrifugal conducting path. Now, since the spinal mechanism with its attached nerves and muscles acts in subordination to the cortical centre, it may be called the *executive* department ; and since the cortical motor centre with its attached conducting path gives the final co-ordination to the cortical impulses before they are transmitted to the spinal nuclei, it may be called the *emissive* department of speech. If the sensory function of speech be now subjected to analysis it will be found that a certain amount of information about the external world may be communicated by one person to another through each and all of our senses. The senses of taste and smell, however, do not take any prominent part in the appreciation of language. Of all the senses the sense of hearing plays the most important part in the sensory function of language, as the vocal apparatus plays in the motor function. It is in connection with the vocal apparatus on the one hand and the sense of hearing on the other that speech is first organised in the individual, and this statement is doubtless equally true with regard to the race. Written and pantomimic speech are appreciated through the eye, and the deaf and dumb may even be taught to appreciate vocal or articulate speech through the same organ. Written speech may, by means of the device of raised letters, be appreciated through the sense of touch, and every person acquires a considerable

4

knowledge of the activity of the motor part of his own speech through
the nerves of muscular sense. The nervous mechanisms which preside
over the sensory function of speech consist of the various peripheral
sense organs with the centripetal conducting paths connecting them
with the highest nerve centres, and of the cortical centres themselves.
The function of the peripheral sense organs with their centripetal
conducting paths may be called the *receptive* department of the sensory
function of speech, but how to name the function of the cortical centres
themselves requires a good deal of consideration. In another place
I have named it the *regulative* department of the sensory function of
language, because the impulses conducted inwards through the various
senses are reduced to order in the cortical centres before they pass
outwards through the emissive to the executive apparatus of speech.
This name, however, hardly gives us any idea of the function of the
cortical centres in reference to language. The peripheral centripetal
apparatus terminates in the cortical sensory centres, but the mental
appreciation of the unit of speech or of a proposition must have for its
correlative an excitation of a much larger portion of the cortex than an
individual sensory centre, and consequently *pyscho-sensory* would not be
a good name for the function of the cortical centres in language. To call
the function the *perceptive* function would answer for some cases, but
not for all. If I were to say "This orange is yellow," I utter a
proposition which the listener could verify by the exercise of his
perceptive faculties, but if I say "The earth revolves round the
sun," the proposition is one which could only be verified by a
prolonged and complicated process of observation and reasoning. The
faculty for the appreciation of complex propositions of this kind, and
even of simple propositions, may be named the apperceptive faculty
in the wide sense in which Kant[1] used the term apperception and its
cognates. The function of the cortical centres lying between the termination of the sensory conducting paths and the emissive part of the
motor mechanism may therefore be named the *apperceptive* department
of the sensory function of speech. The apperceptive faculty is susceptible
of being analysed into simpler elements; but in the meantime we shall
speak of it as if it were a simple and homogeneous faculty. Our analysis
has, therefore, led us to recognise that speech consists of a motor and of
a sensory function, and that the former of these may be divided into an

___

[1] "Verbindung liegt aber nicht in den Gegenständen und kann von ihnen nicht etwa durch
Wahrnehmung entlehnt und in den Verstand dadurch allererst aufgenommem werden, sondern
ist allein eine Verrichtung des Verstandes, der selbst nichts weiter ist, als das Vermögen, *a priori*
zu verbinden und das Mannigfaltige gegebener Vorstellungen unter die Einheit der Apperception
zu bringen, welcher Grundsatz der oberste im ganzen menschlichen Erkenntniss ist."—Kant's
Kritik der reinen Vernunft, Herausg. von G. Hartenstein, Leipzig, 1868, p. 117.

5

emissive and an *executive* department; and the latter into a *receptive* and an *apperceptive* department, with their corresponding nervous mechanisms. Now the term aphasia is not meant to include disease of all these functions and mechanisms. Disease of the executive department of the motor mechanism of speech gives rise to the symptoms known as bulbar paralysis, or to local diseases of the nerves and muscles by which thoughts are expressed, but these diseases ought to be excluded from the definition of aphasia. And again, disease of the receptive department of the sensory mechanism gives rise to deafness, deaf-mutism, blindness, anæsthesia, and various other sensory disorders ; but those symptoms also ought to be excluded from any definition of aphasia. Again, every kind of disorder of the higher mechanisms of speech is not an aphasia. The epileptic cry is probably caused by a nervous discharge from the emissive department of the motor mechanism of speech, but this cry is not regarded as an aphasia. The disorder of the nervous mechanism which constitutes an aphasia must be of the nature of a paralysis, although the paralysis need not necessarily be complete. Aphasia may, therefore, be used as a generic term to include diminution or loss of the function of the emissive department of the motor, or of the apperceptive department of the sensory function of speech, or of both these departments combined. In accordance with this definition, aphasia may be divided into (1) motor aphasia, (2) sensory aphasia, and (3) combined motor and sensory aphasia.

1.—MOTOR APHASIA.

In cases of complete motor aphasia the patient is unable to communicate his thoughts by articulate words, by writing, or by pantomime. Certain terms have been employed to designate each of these conditions. Loss of the power of communicating thought by articulate words has been named *aphemia* (ἀ priv., and φημί I speak), by writing *agraphia* (ἀ and γράφω I write), by pantomime *amimia* (ἀ and μῖμος an imitator), and by all these methods of expression at once *asemasia* (ἀ and σημαίνω to show by a sign). Loss of the power of expressing thoughts has also been named *ataxic* aphasia.

It is important to remember that a patient may be the subject of a complete motor aphasia, and yet be able to utter some words. In Dr. Hughlings-Jackson's language the subject is *speechless* but not *wordless*.[1] The words which the patient is able to utter continue, as a rule, the same in the same patient, and have therefore been named "recurring

[1] Hughlings-Jackson (J.). "On affections of speech from diseases of the brain."—*Brain*, Vol. I., 1878-79, p. 313, and Vol. II., 1879-80, p. 204.

utterances." Utterances of this kind, like "yes" or "no," which are repeated on all occasions, whether appropriate or not, do not possess any intellectual value. In addition to the usual recurring utterances of "yes" or "no" patients sometimes repeat such phrases as "Come on to me," or "I want protection."[1] The man who kept repeating the phrase, "Come on to me," was a railway signalman, who had been taken ill on the rails in front of his box; while the man who could only say, "I want protection," had his head injured in a brawl. Dr. Hughlings-Jackson makes the very probable supposition that in these cases the frequently repeated phrase were the last words spoken, or which were in a state of mental preparation for utterance, when the damage occurred to the brain. It is not improbable that words spoken, or about to be spoken during a period of great excitement, will leave permanent traces on the organisation of the brain, which renders them liable to be subsequently uttered as interjectional phrases during emotional states. A patient may also use when excited words and phrases, such as "God bless me," or profane oaths, in an interjectional sense; but it is evident that these words are expressive of the feelings or emotions, and not of the cognitions or thoughts. A patient may also utter a phrase which is appropriate to the surrounding circumstances, such as "good-bye," when a friend is leaving.[2] Words of this kind have been frequently repeated in the previous experience of the individual, and always under more or less similar circumstances, and have thus become so deeply organised that they possess little more value than a reflex act. The fact that a patient who may say "good-bye," on taking leave of a friend, and is quite unable to repeat the words when asked to do so, shows that the utterance is little more than a reflex act without any intellectual value.

When the aphemia, or the loss of articulate speech, is not complete, the patient retains the full use of simple words like "yes" and "no," and even when he says "no" to express assent as well as dissent, if he is able to indicate by the aid of pantomime in what sense he wishes the word to be understood, this shows that the word possesses an intellectual value. Although all forms of intellectual expression are often simultaneously affected in motor aphasia, yet this is by no means always the case. A man may be totally unable to express his thoughts by articulate words, yet be able to write with tolerable freedom, and to express his wants by pantomime. Loss of the power of written speech while

<hr/>

[1] Hughlings-Jackson (J.). "Remarks on the occasional utterances of speechless patients."— *The Lancet*, Vol. II., 1867, p. 70.

[2] Broadbent (W. H.). "A case of peculiar affection of speech, with commentary."—*Brain*, Vol. I., 1878-9, p. 494.

articulate speech is retained or re-acquired is rare, though cases of the kind have been reported; but a case of loss of intellectual pantomime with retention of the power of articulate speech and of writing is not known to me. But the functions of language which the subject of motor aphasia retains are just as remarkable and as noteworthy as those he loses. The patient is able to perform all the general movements of the tongue and lips, although he may be totally unable to articulate a single word. There is no difficulty of deglutition, dribbling of saliva, or any of the usual symptoms of labio-glosso-pharyngeal paralysis; in other words the executive department of the motor function of speech is quite healthy. The language of the emotions is little or not at all affected. The patient can smile, frown, laugh, or sing, and the few words he can use, such as "yes" or "no," may be uttered with such variations of tone and gesture as to indicate when he is angry or joyful, these words being then, in Mr. Herbert Spencer's language, more akin to song than to speech.[1] The sensory function of speech is also quite unaffected. The patient understands all that is said to him, and remembers what is said to him or what he reads himself. He will point to surrounding objects, and recognise drawings of them, provided they were known to him before his illness. He also recognises handwriting, and is able to play at cards and other games. The following case is an example of motor aphasia, although an attentive scrutiny shows that the motor disability is accompanied by a slight degree of sensory aphasia.

Case 1.—W. D., aged 24 years, was admitted to the Manchester Royal Infirmary, in June, 1884, under the care of Dr. Ross. The following history was obtained from his wife. She states that the patient, who was an out-door worker on the railway, was always a strong and healthy man, and never suffered from any serious illness up to the date of the present attack. She does not think that he ever suffered from rheumatic fever, and there is no evidence of syphilis. About twelve months ago, while loading a waggon at the Salford Station, he was seen to turn suddenly round and to fall. He was carried home in an unconscious condition, and it was soon observed that he was completely paralysed on the right side of the body. For the first two months he did not seem to understand anything that was said to him, and did not take much notice of surrounding objects and events; but at the end of that time he began to take an intelligent interest in what was passing around him, and by promptly obeying ordinary requests, such as "put your tongue out," "sit up in bed," &c., he showed that he understood much of what was said to him. He also appeared to

[1] Spencer (Herbert). "The origin and function of music." Essays—Scientific, political, and speculative, Vol. I., 1868, p. 221.

take an interest in looking over a newspaper. He continued, however, quite unable to utter a single word. He gradually recovered from his paralysis, so that in two or three months he was able to walk about on level ground, although he dragged the right leg, and did not use the right hand much.

Present Condition.—The patient is a short, stiff, and muscular man, and looks very healthy. On close inspection it is seen that his lips are of a rather blueish colour, and his face is somewhat congested. Both hands are of a reddish-blue colour, and always very cold, these phenomena being specially well-marked in the right hand. The pulse is very feeble, but regular, and beats 72 in the minute. On examining the chest it is seen that the apex of the heart beats two inches below the nipple, and a little to the inside of the mammary line. The deep cardiac dulness reaches as high as the upper border of the third rib in the left parasternal line, but the deep cardiac dulness does not extend to more than half-an-inch to the right of the right edge of the sternum. On auscultation a soft systolic murmur is heard at a localised spot over the apex of the heart, but it is replaced by an impure first sound at mid-axilla. The second sound is accentuated over the third left costal cartilage. All the other internal organs are healthy. In walking the patient drags the right leg a little, and the patellar-tendon reaction is increased in it, and a little ankle clonus can be elicited by careful manipulation. The grasp of the right hand is fairly good, although weaker than that of the left, and the tendon reactions at the wrist and elbow are increased on the right side. The face is drawn slightly to the left side.

The patient is unable to utter a single word either in answer to a question or by way of repeating what is said. Asked his name, he shakes his head mournfully to indicate his inability to reply. Asked if his name is William, he nods to give assent. Asked if he is one hundred years of age, he laughs and gives a short grunt which is something like an attempt to say " No." He does not appear to have any kind of recurring utterance. He can write down his own name in a fair hand, but can only write the first half of Manchester when asked to do so, although he can copy the name in full by frequent reference to the printed page. On being requested to show by signs how he would ask a second person to close the door, he laughed and raised his hand and significantly pointed to the door. The patient can protrude his tongue with a scarcely perceptible deviation to the right. He can roll his tongue into a tube and turn the tip upwards towards the nose. He can also blow out his cheeks. There is no dribbling of saliva, and no difficulty of deglutition. He obeys promptly any ordinary request in articulate speech, such as " Put out your tongue," "Close your eyes," "Show your hand," &c. When shown a simple request, such as "Put out your tongue," in writing, he cannot be got to obey it, but his wife says that he is fond of looking over the newspaper, and she believes that he understands it, as he sometimes points out to her a paragraph he must know would interest her. His sight, hearing, and all his other senses are normal.

December 16.—The patient has been under treatment as an out-patient in the Infirmary for upwards of six months, but his condition remains almost unchanged. He has now the full use of "Yes" and "No," and always uses the words correctly. When asked his name, he says it is "Bill." He has also been taught to repeat most of the letters of the alphabet. The explosive consonants were the first letters acquired by him. He generally keeps his lips slightly apart. I first got him to look steadily at my face, and, closing my lips firmly, I touched his lips and pointed to him to do the same. When he closed his lips I then pronounced the letter p, and he immediately followed with a b. He smiled and seemed pleased. After some exercise we got him to pro-nounce many of the letters of the alphabet, and a few monosyllabic words like man and cow, but he has not been systematically exercised, and his vocabulary, even for the repetition of words, is still very limited. To-day he was asked, "Close your eyes," and he immediately obeyed. The same request was now made in writing, and it was instantly obeyed. I now wrote down "Put your tongue out," and he promptly complied with this request likewise, this being the first time that I was able to secure obedience to a written command. In other respects the patient is about the same as he was when he came first under observation.

Remarks.—When this patient was first seen in June he was the subject of a complete aphemia. The agraphia was not total, inasmuch as he was able to write his name and to copy simple words, while his pantomimic speech was apparently not much impaired; in other words there was little or no amimia. No defect could be discovered in the sensory side of speech through the ear, but there was a suspicion of the existence of some degree of word-blindness, inasmuch as he could not be got to obey a written request. When examined in December, the word-blindness, if it previously existed, had disappeared. It is, however, important to observe that during the first two months after his attack he appears to have suffered from a more or less complete sensory, as well as motor, aphasia. The patient's wife stated emphatically that during the first two months of his disease he was quite unable to understand anything that was said to him. An opportunity was offered to me to observe this condition of combined sensory and motor aphasia about twelve months ago, in a boy, who was at first under the care of my colleague, Dr. Morgan, and was subsequently under my own care for many weeks in the Barnes Convalescent Hospital at Cheadle. This boy, who, like W. D., had a soft systolic murmur over the apex of the heart, was suddenly seized with an apoplectic attack and right hemiplegia. On entering the Infirmary, soon after the attack, he was found to be suffering from right-sided hemiplegia and complete sensory and motor aphasia. He was not only unable to utter a single word, but he also took no heed of any ordinary request, whether addressed to him by means of spoken or written language. At the end of about three weeks he recovered on the sensory side so far as to put out his tongue on being requested to do so. In a few weeks longer he recovered so far as to obey all ordinary requests, and to take an interest in a lesson book, so that the sensory aphasia may be said to have completely disappeared; but he was still unable to utter a single word. In the meantime the paralysed limbs had recovered their motor power, and scarcely a trace of hemiplegia could be discovered, with the exception of some excess of the tendon reactions at the right wrist and knee. During the whole of this time the patient was unable to utter a single word, a short grunt with a nod for assent, and a shake of the head for dissent, forming the whole of the expressive part of his speech. At the end of about five weeks from the commencement of the attack he was transferred to the Convalescent Hospital at Cheadle, and under the tuition of Dr. Wansbrough Jones and myself he soon learnt to repeat the alphabet, the explosive consonants being first taught him. In a short time he acquired a pretty copious vocabulary of monosyllabic words, such as man, horse, cow, but he was only able to say most of his words by repetition, and was still unable to construct a sentence. He was kept for about two months in the hospital, and on being discharged was lost sight of. This case shows that the early stage of a motor aphasia is sometimes, at least, accompanied by a sensory aphasia, and that after some weeks recovery takes place from the sensory disability, leaving a more or less pure motor aphasia. A combined sensory and motor aphasia is probably always present in all cases in which the loss of speech, if it be persistent, is accompanied by a decided hemiplegia, and it is in those cases alone in which the motor aphasia is accompanied by a slight degree of facial paralysis, without the limbs being much implicated, that a pure motor aphasia is present from the commencement of the attack. In the case of W—— D——, the loss of the power of writing was not quite so profound as that of the power of articulate speech; and in other cases the disproportion between the interference with these two functions is still greater than it was in this one. In a case I saw some years ago with Mr. Sutcliffe, of Stretford Road, the patient was quite unable to articulate a single word, but he could write with great readiness pertinent replies, and give instructions about his business on a slate; and although the writing was not very good, Mr. Sutcliffe could decipher it without much difficulty.

Cases in which the power of speech is retained or soon re-acquired, while the power of writing is lost, are not frequent, but an instance of this kind has been reported by Pitres.[1] The subject, a merchant, aged 31 years, who had contracted a chancre ten years previously, was seized with an apoplectiform attack, right-sided hemiplegia, and a certain degree of aphasia. For some weeks after the attack the patient became gradually worse and more comatose, but after energetic anti-syphilitic treatment improvement set in, and the patient regained consciousness, the paralysis gradually disappeared, and the power of articulate speech returned. The patient was examined by Pitres eighteen months from the beginning of the attack, and only slight traces of the previous hemiplegia could then be discovered, while he could talk freely, and without any evidence of defective articulation or other disorder of spoken speech. The right hand was redder and colder than the left, and it was often covered with chilblains; the sensibility to contact and pricking was nearly the same in both hands, but with careful testing the muscular sense was found to be slightly, and only slightly, diminished in the right as compared with the left hand. The only other disorder of sensibility that could be discovered was a right-sided hemianopsia. All the general movements of the right hand were made with precision, and the patient could read aloud without hesitation. He was, however, quite unable to write a single word or even a single letter to dictation with the right hand, although he had acquired the power of writing fairly well with the left hand. When asked to copy a printed word he began slowly, and after looking frequently at the page he could reproduce the word in printed but not in written characters, and when a written word was presented to him he could reproduce it slowly by copying it letter for letter in written characters, but when the model was withdrawn he was quite unable to write the word in written or printed characters. With numbers it was the same; the patient could read them without hesitation, and could add, subtract, and make other calculations like any healthy, educated person; he could also copy slowly with the right hand a number presented to him, whether printed or written, but he was quite unable to write numbers spontaneously or to dictation. The patient

[1] Pitres (A.). "Considérations sur l'agraphie à propos d'une observation nouvelle d'agraphie motrice pure."—*Revue de Médecine*, Nov., 1884, p. 855.

could, however, copy with the right hand geometrical figures, such as a circle or triangle, and he was also able to draw with it a well-proportioned outline of the human figure. There can be little room for doubt that in this case the damage to the brain had occurred on the motor side, and yet, although there was almost complete motor agraphia, the patient had recovered the full use of articulate speech. These remarks on motor aphasia must suffice, and we shall now proceed to discuss the various forms of sensory aphasia.

2.—SENSORY APHASIA.

In sensory aphasia the chief disorder of speech is to be found in what we have called the *apperceptive* faculty. The slightest consideration of the mechanism of speech will, however, render it manifest that the disability will not be limited to this faculty, but that all the faculties, the nervous mechanisms of which lie in front of the main lesion, will be thrown into disorder. During the activity of the collective speech mechanisms the currents pass, speaking broadly, from the ear to the cortex of the brain, and from the cortex to the articulatory organs in spoken speech, and from the eye to the cortex of the brain, and from the cortex to the hand in written speech. Interruption of continuity of any portion of these routes will damage not only the function of the local part injured, but will throw into disorder all the parts which lie anterior to the lesion. If a person becomes absolutely deaf in early infancy, he is afterwards, unless subjected to a very special education, mute also; the sensory disorder has prevented the development of the motor apparatus. Similarly, a person born blind is incapable of acquiring the power of writing by any ordinary training; here also the sensory disorder has placed a bar to the development of the motor capacity. These illustrations will serve to render it clear that in cases of pure sensory aphasia it may be expected that a motor disorder of speech will also be present. In sensory aphasia, therefore, and, indeed, in every form of aphasia, the most valuable indications as to the particular part of the nervous mechanism of speech which is injured is to be found, not in observing the peculiar kind of motor disorder which is present, but in noting how the patient comports himself when certain requests or questions are addressed to him through the ear or eye; that is, in spoken or written language respectively. If a patient takes no notice of whistling or other loud noises in his vicinity it is concluded that he is deaf, but if he turns round and scans with intelligent expression the direction from which the sound has emanated, it is concluded that there is no insuperable barrier to the passage of impulses from the local apparatus in the ear to the cortex of the brain. But if the patient who gives distinct evidence of hearing ordinary noises fails to obey simple spoken requests, such as "Close your eyes," and if his other conduct shows that his not complying with the request is due to simple perversity, it will be safe to conclude that he has not appreciated the nature of the request, or in other words, that the part of the apperceptive faculty of speech in relation with the sense of hearing is disordered. In the same way, if an educated man avoids obstacles, and takes notice of surrounding objects, yet fails to obey simple written or printed requests, or is unable to read, while still capable of expressing his thoughts correctly in spoken words, then it is concluded that the part of the apperceptive faculty which is in relation with the sense of vision is disordered. In such cases a very valuable test, first proposed by Dr. Hughlings Jackson, is to ask the patient some ridiculous question, such as "Are you a hundred years of age?" If the question is understood, it is sure to evoke an emphatic denial on the part of the patient, accompanied by lively evidences of amusement or indignation. But instead of dwelling further upon the tests which are to be applied in cases of sensory aphasia, we shall proceed to show them in operation by describing cases illustrative of the different forms of this disorder of speech. The simplest varieties of sensory aphasia are afforded by the cases, which were first clearly recognised by Wernicke, and which have been named by Kussmaul *word blindness* and *word deafness*. Pure examples of these affections are rare. In practice they are most frequently met with variously combined in the same patient.

The following case is a very good example of uncomplicated word-blindness, the disorder of speech being accompanied by a well defined right-sided bilateral homonymous hemianopsia.

Case 2.—Robert Marshall, aged 57 years, was admitted into the Manchester Royal Infirmary under my care on July 21st, 1884.

Previous History.—The following account of the patient was obtained partly from himself and partly from his wife. His occupation was that of engraver on copper, and he was a very well educated man, and fond of reading. He never suffered from exposure or any particular hardships, and he was always very temperate and regular in his habits. There is nothing in his history pointing to syphilis. He has suffered for many years from deafness, and his wife thinks that he was as deaf as he is at present for the last two or three years. For some years he has had to micturate frequently, and the quantity of urine passed has been much more copious than in former years. The patient states that he has never suffered much from thirst, but he acknowledges that he drinks a considerable quantity of fluid in the twenty-four hours. With these exceptions he has not complained of any symptom worth noting up to within five months of his admission to the Infirmary. At that time he was working very hard on a Saturday afternoon when he was seized with dizziness and a general feeling of illness. Finding himself unable to continue his work he lay down and slept, but after a short time he awoke suffering from a violent headache, and he found that he was unable to speak clearly, and could not define objects. On the following day he was worse rather than better ; he experienced considerable difficulty in expressing his wants. His wife says that he did not give pertinent replies to her questions, while he was unable to see clearly, objects appearing to float before him. Both he and his wife state emphatically that at no time was there any sign of paralysis. He states that his power of speech returned at the end of about a week or ten days, but he does not believe that his sight has in any way improved since the attack.

Present Condition.—R. M. was in the Infirmary for upwards of five weeks, and during that time the following notes were taken. The patient is a tall and stout man, weighs over fifteen stones, and looks well nourished and healthy. The urine is free from albumen, but it contains a considerable amount of sugar, the quantity as estimated by the fermentation test varying from 25 to 30 grains per ounce. The quantity of urine passed in the twenty-four hours has varied from 20 to 80 ounces, the average being about 50 ounces. The apex of the heart is displaced slightly downwards and outwards, and the second sound at the base is highly accentuated and of metallic quality. The arteries at the wrist feel somewhat unyielding and inelastic, but they are not calcareous. The patient is so deaf that he only hears the ticking of a watch on either side when it is brought in contact with the ear. His sight enables him to avoid obstacles in walking, but when he looks at small objects he has to use double convex glasses, as he is the subject of a considerable degree of presbyopia. He is, however, suffering from a right-sided bilateral homonymous hemianopsia, the completeness of which is testified by the accompanying charts taken on July 29th, by means of Dr. McHardy's perimeter (Figs. 1 and 2). The patient's deafness has rendered him stupid and difficult to examine, but he converses sensibly and freely about himself, and especially about his visual defect, of which he complains most. He also names any objects presented to him correctly, and gives rational replies to all questions addressed to him. In short, during his residence of upwards of five weeks in the Infirmary, the fact that he had any special disability of speech had completely escaped the attention of myself and several other very competent observers, who were led to examine him more or less carefully.   He was discharged on August 28th, and made an out-patient.

FIG. 1.—LEFT.

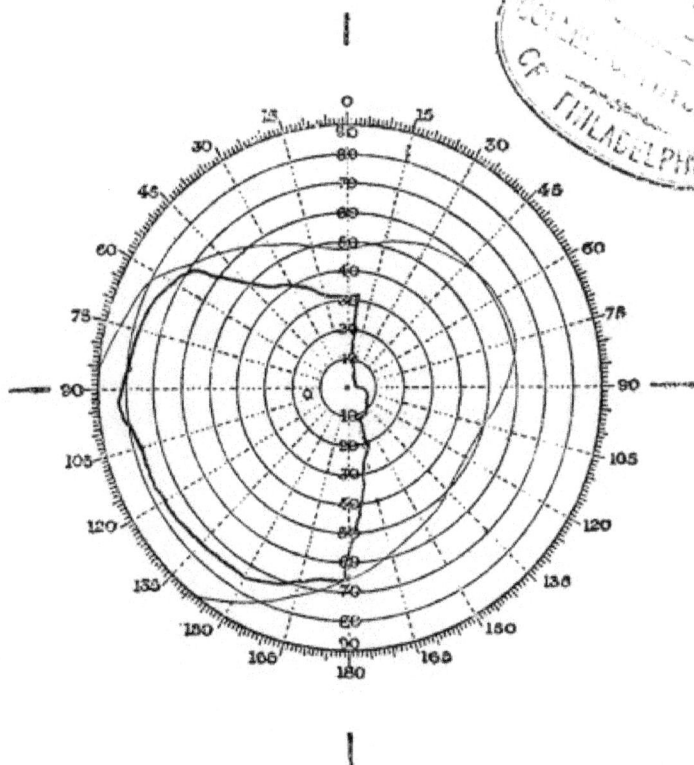

Figs. 1 and 2.—Charts of the Fields of Vision in the case of Robert Marshall, showing by the heavy line bilateral homonymous hemianopsia.

December 16, 1885.—R. M. has attended pretty regularly as an out-patient at the Infirmary since the date of the last report, except for a month during last summer, when he was a patient in the Convalescent Hospital, at Cheadle, where Dr. Wansbrough Jones and myself had frequent opportunities of examining him together.   The first time that any disorder of speech was observed to be present was one morning soon after he became an out-patient.   I was demonstrating the presence of hemianopsia to some students, when he volunteered the following statement: "My sight is very curious.   I cannot see things near me so well as those at a distance; I cannot, for instance, see to read this"— pointing to the heading of "Manchester Royal Infirmary," in capital letters, on the prescription book for out-patients which he held in his

left hand—"as well as those out there," pointing to signboards on which
the names of various mercantile firms are painted in large gilt letters,

FIG. 2.—RIGHT.

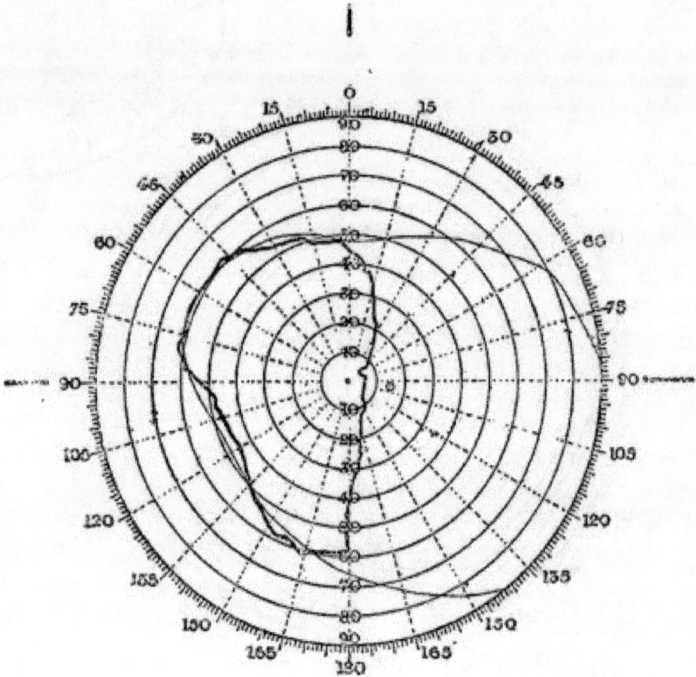

and which would be about one hundred yards distant from us. I then
directed his attention to one of those signboards, and asked him to tell
me the name on it. After looking at it for some time earnestly through
his spectacles, he said, "Thomson Brothers I make of it; I do not know
whether I am right." No such name, however, was anywhere to be
seen. I tried him with another, and another, and still I got the same
reply, "Thomson Brothers." On being asked to spell the name on one
of the signboards, he began, "T, h, o, m," &c., and on being told that he
was wrong he began with some other letters, not one of which corresponded with the name to which his attention was directed, or with any
other visible name. I then got him to write down "Thomson Brothers,"
which he did in a somewhat tremulous but very readable hand. I now
requested him to write his own name. "Do you mean with my eyes
open or my eyes closed?" he demanded. "With your eyes open, of

course," I said. His wife interposing, said, "He can write much better with his eyes closed than he can when they are open." He now wrote, with open eyes, somewhat hesitatingly, but without mistake, "Robert Marshall," but on closing his eyes he wrote his name in a much firmer hand, and with less hesitation, than he did while using his sight. At my request he wrote down "Manchester." The paper on which he had been writing was now withdrawn from him, and presented again after a short interval. Pointing to the first line written by him, he read "Thomson Brothers," which was correct; but the second line, which was his own name, the third, which was "Manchester," and the fourth, which was his own name again, were read equally with the first as "Thomson Brothers." On being tried to spell from a printed page, every letter was found to be wrong, but singularly he could read aright two or three out of the nine single numbers, the number 4 being generally correct. The following note was now handed to me by one of the patients, who was the subject of locomotor ataxia, with white atrophy of the optic discs. It was written on a memorandum form, headed "Henshaw's Blind Asylum," and proceeded: "Dear Sir,—I shall be much obliged if you let me know whether or not you consider it likely that A. B. will recover his sight.—Yours, &c." I handed this note to Marshall, and asked him to read it. He immediately took it, scanned it carefully through his spectacles, moving it backwards and forwards a little until he got it to the best distance for his sight, and he then read slowly and deliberately, but without much hesitation : "Manchester Royal Infirmary. Dear Sir,—You are requested to bring this note with you the next time you come to the Infirmary," and ended up with his usual formula, "that is what I make of it, I don't know whether it is right or not." His wife then said : "At home he pretends to read the newspaper, and he reads such stuff, all made out of his own head. Last night he was reading to me aloud, and it was all stuff made up, you know. After a time he got tired, and asked me to read. I read it as it was in the paper; it was something about the war (Egyptian). He was very quiet for some time, and then he asked, 'Is that what it says in the paper?' and when I told him it was, he said, 'Well, then, I must be an idiot.'" She then added, "He often says he thinks he must be an idiot now."

In August, 1885, Dr. Wansbrough Jones and myself, in testing him, gave him the following paragraph to read : "The Bishop of Manchester.— It is stated that the Bishop of Manchester is slightly indisposed, and will be unable to fulfil his engagement to preach at St. Paul's, Brunswick Street, next Sunday morning." After taking considerable time and care in adjusting his spectacles, and getting the point of the index finger of his right hand on the paragraph, he read : "The money market has

B

been brought to a close, and considered much more easy than it was, and I think it is going to improve."

December 16.—R. M. was again examined at the Infirmary to-day, and he is found to be practically unchanged. The hemianopsia appears to be as complete on both sides as it was when the patient was examined about eighteen months ago. When asked to read he makes very elaborate preparations, and utters a few sentences which have not the remotest connection with anything that is before him on the printed page. He can write a short word to dictation, but cannot complete polysyllabic words or a sentence. Asked to write a word of two or three syllables, he may, by writing very fast, write it correctly; but, if he once raises his pen, he gets confused, and soon gives up the attempt. After raising his pen in the middle of a word, he generally makes several attempts to join the remainder of the word to the last letter, and, not succeeding in this, he begins the word anew, but after one failure he seldom succeeds in completing it. Asked to write "command," for example, he wrote "comma." He then raised his pen, got confused, and, after several futile attempts to complete the word, he began anew and wrote "com," when he got confused a second time and gave up the attempt altogether. But if, when he stops writing in the middle of a word, the next succeeding letters are loudly sounded in his ear, he can often complete even a long word. He generally recognises his own name, and always reads correctly "Manchester Royal Infirmary" as the heading of his prescription book. The next line in the book, "Dr. Ross's patient," is sometimes said to be "Dr. Ross and company." It would, however, appear that his recognition of these words is more or less guesswork, inasmuch as he names other words "Manchester Royal Infirmary," as he did in the case of the letter from "Henshaw's Blind Asylum." He seldom or ever declines to give some name to a word presented to him, but beyond those I have just mentioned I have never known him to have been correct. He is even unable to recognise a single letter. Occasionally he names a letter correctly, but the next time it is presented to him he generally names it wrongly, thus showing that his being right was a mere accident. With numbers he can generally recognise 3, 4, and 7, but with the remaining numbers he is more often wrong than right. He recognises quickly and names geometrical figures, such as a square or a circle, and recognises portraits. I showed him, for example, a photograph of myself, and asked him whom it represented. On getting it at the right focal distance for his sight, he instantly replied, " It is not a very good one, but it should be Dr. Ross." His spoken language is not at all impaired. His deafness makes him a stupid man to converse with, but there is no recognisable defect of any kind in his spoken speech. He names correctly and promptly all objects presented to him, and seems much amused at being asked the names of his bodily organs. He obeys, without any hesitation or confusion, all spoken requests, and gives rational replies to all questions.

January 16, 1886.—Mrs. Marshall came to the Infirmary this morning for her husband's medicine—an iron mixture. In speaking of him she volunteered a very interesting piece of information. " He has given over attempting to read the newspaper now," she said. " A short time ago," she continued, " I hardly knew what to do with him. He tried to read the newspaper, and he kept saying, ' I don't know what is the matter with the papers now-a-days, they are filled with such silly stuff.' I frequently had to get him as many as four newspapers in a day to see if it would satisfy him. At last I was obliged to tell him that it was not the newspapers that were wrong, but that it was he who could not read, and now he has given over trying."

In the following case the word-blindness was at first accompanied by a considerable degree of word-deafness, but the patient is recovering from the latter condition, while the former persists. The speech disorder is accompanied by an incomplete homonymous hemianopsia. For the notes of the case I am indebted to Mr. Windle, my clinical clerk.

Case 3.—James Proffit, aged 53 years, was admitted as an in-patient to the Manchester Royal Infirmary, on October 7th, 1885.

History.—The following account was obtained chiefly from the patient's wife. They have been married for about 25 years. Twelve months after they were married his wife had a stillborn child, which she thinks was born at the seventh month. Two other stillborn children followed in succession at intervals of twelve months; then she had a living child at full time, but it died of convulsions at three weeks old, and another stillborn child followed at the lapse of another twelve months. After this time she has had five living children, all of whom are now alive and healthy. The patient does not own to having contracted syphilis when a young man. The patient has always been a very healthy man, and his occupation—an outdoor labourer—has been, on the whole, a healthy one, although it exposed him to alterations of weather. He has never been a hard drinker, but at times took a moderate quantity of beer. A month previous to the date of his admission, he was suddenly seized, when getting up in the morning, with some kind of attack, of which neither he nor his wife can give a very satisfactory account further than that, as he recovered from the first loss of consciousness, he was found to be speechless, and paralysed on the right side of the body. For the first few days he could hardly understand anything that was said to him, and his attempts at speaking were almost unintelligible, but he soon showed signs of amendment, and he began to understand better, and to say a few words, which sufficed to make known his ordinary wants.

Present Condition.—On admission, the patient is found to be a short, strong, muscular man, with a tendency to obesity. The area of cardiac dulness is slightly enlarged downwards and to the left, the second sound at the base is considerably accentuated, and the arteries at the temples and wrists are somewhat tortuous, but not knotty or calcified. The urine is free from sugar or albumen, and, with the exception of some bronchitis and emphysema, the thorax and abdominal organs are free from disease. The patient drags the right leg a little, and there is a slight relative increase in it of the patellar-tendon reaction. The right hand is paralysed, and the fingers are flexed into the palm, and rigid, but the patient can move that extremity pretty freely at the elbow and shoulder joints. The tendon reactions are exaggerated at the right wrist and elbow. There is a slight relative loss of expression of the right side of the face, but so slight as to be scarcely perceptible, but there is no deviation of the tongue on protusion. On admission, the patient had complete right-sided bilateral homonymous hemianopsia. He had a confused and lachrymose appearance, and when asked his name he drawled out " Ja-a-ames Pro-o-ofit," prolonging the open vowel in each word to a most comical degree. He is unable to give a connected account of himself, but, after manifesting some confusion, and making some mistakes, he ultimately comes to interpret and obey correctly such simple requests as, " Put out your tongue ; " " Close your eyes ; " " Show your hand." To the question, " How old are you ? " he replies, " I ca-a-a-na te-e-ll ; " but when asked, " Are you a hundred years old ? " he rejoins, with manifest surprise, " Tut ! No-o-a, ma-a-an. I-i-i'm o-o-only a yo-o-oung fe-e-llow yet." Asked if he is 50 years of age, he says, " Mo-o-re," and when 53 is mentioned, he replies, with evident satisfaction, " Tha-a-at's it." Asked the name of his hand, he says, " I c-a-ana just tell, I kno-c-ow it qui-ite we-ell, but I ca-a-ana te-ell." " Is it your head ? " " No-oa, sir." " Is it your foot ? " " No-oa, sir." " Is it your hand ? " " Ye-es, sir, tha-a-at's it," he replies, with the usual manifest air of satisfaction. Tried with his head, eyes, and other bodily organs, and with surrounding objects, it was the same ; he was unable to tell the correct name, but he at once recognised whether a name is correct or not when it was uttered in his hearing. He recognises James Proffit, as written on the bed-ticket, but he is unable to read, or even to spell, my name, or any of the printed words, such as " Manchester Royal Infirmary," on the ticket. He is also unable to make out a single word or letter from a newspaper. He is incapable of identifying written or printed numbers. He can, however, tell whether a particular written or printed letter or number is or is not correctly named in his hearing. Any short and simple word uttered in his hearing he is able to repeat at once, but gets confused when asked to pronounce long words with many syllables such as " Constantinople,"

FIG. 3.—LEFT.

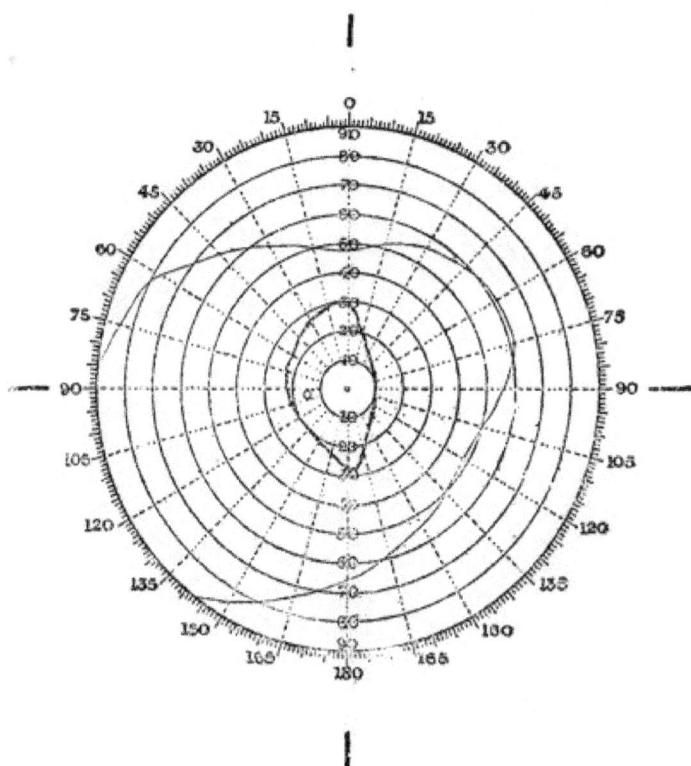

Figs. 3 and 4.—Charts of the Fields of Vision in the case of James Proffit showing by the heavy line general restriction of the fields with incomplete bilateral hemianopsia.

and often sticks at the end of the second syllable. Asked to write his name, he takes the pen in his left hand, his right being paralysed, and makes up and down strokes like a continuous letter m, but he cannot write James, which he is evidently attempting. He also writes from right to left, and slants the strokes from left to right, as in mirror-writing, this being the opposite way to the usual method of writing with the right hand. After struggling painfully with these up and down strokes for a short time, he throws down the pen and declares his inability to write; and then adds, " I ne-ever wa-as a goo-od scho-o-olard. ' " But you could always write your name quite well?" I said enquiringly. "Oh ye-e-s, sir." " And you could read the newspaper." "Oh ye-es, sir; qui-ite we-ell." The fact that the patient's wife had several still-born children rendered syphilis probable, notwithstanding his denial of having had a primary sore; and he was consequently ordered ten grains of iodide of potassium three times a day, and two grains of blue pill night and morning.

FIG. 4.—RIGHT.

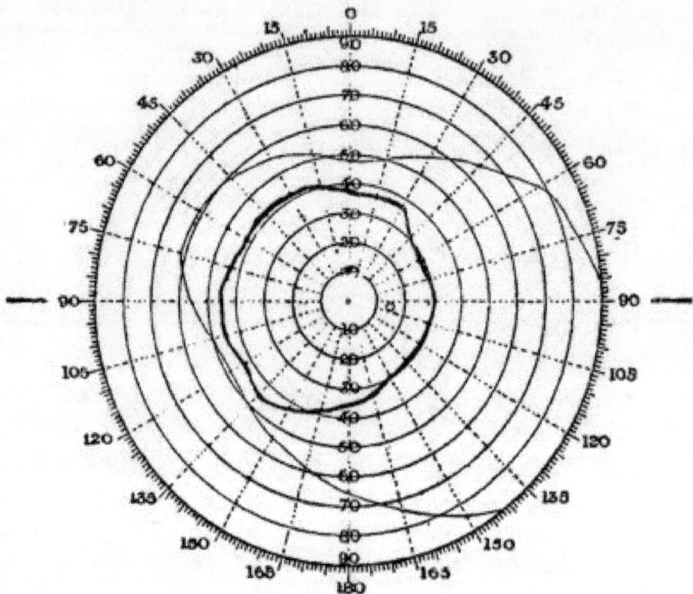

Nov. 19.—During his residence in the Infirmary the patient improved greatly. Almost every trace of hemiplegia disappeared from the right side of the body, even the tendon reactions at the knee and wrist having become almost normal. Soon after his admission it was observed that his fields of vision were rapidly enlarging on the blind sides, and the annexed perimetric tracings (Figs. 3 and 4) which were taken on October 19— twelve days after admission—show that the bilateral hemianopsia was by no means so perfect at that date as it was when the patient was first examined. At the date of the present report the fields have enlarged still further, and instead of a complete bilateral hemianopsia the fields are

only slightly restricted on the right side. The patient can now name the different parts of his body and the common objects by which he is surrounded, but still hesitates over the name of any unaccustomed object presented to him. The power of naming has, however, come to him only gradually and by a certain amount of training. He experienced a good deal of difficulty in giving a name to his own ear. When the word "ear" was uttered in his hearing, he at once recognised that it was the appropriate name for his organ of hearing, and repeated the word readily, while holding the lobule of his ear between his finger and thumb. If asked the name again after a short interval, he would say "There, I've lost it; I cannot tell," and he would then close his eyes and turn his head now to one side then to the other, as if he were intently listening; but all in vain, and he would soon give up the effort. After the lapse of a few days, however, when once the word "ear" was uttered in his hearing, he would retain the use of it for the remainder of that morning, but would forget it by the following day. After another interval of a few days he was able to retain the use of the word permanently. When he had made considerable progress in acquiring the names of his own organs and of the surrounding objects, I presented my watch to him and asked him to name it. He was unable to do so. I then asked "Is it a bell?" to which he promptly replied "No, Sir!" "Is it a clock?" I now asked. To this question he replied somewhat dubiously, "That's something like it"; but when asked "Is it a watch?" his expression brightened up and he immediately responded "That's it— a watch." He was now requested to read off the time from the watch. After several futile efforts he placed the tip of the index finger of the right hand opposite the figure *one*, and carrying it slowly round the dial, he named correctly each figure in succession until he came to *twelve*; but even then he was unable to tell the correct hour, and never attempted to read the minutes. I now asked him if the right time were a quarter past six, purposely naming the wrong time, and he immediately answered in the negative; but on asking if it were twenty-five minutes past eleven—the correct time—he said "That's more like it"; but his assent was not given with that bright intelligence he manifests when he recognises the forgotten name of an object. He was now discharged from the Infirmary and made an out-patient.

Jan. 16, 1886.—James Proffit appeared as an out-patient at the Infirmary this morning. He has improved considerably in his speech since the last report; there is not much restriction to be discovered now in the fields of vision, and the hemiplegia has almost disappeared. He named correctly most of the objects presented to him, but on being asked the name of one of those table-bells which is rung by striking a brass knob at the top of the instrument, he hesitated, and for some time failed to recollect it. He closed his eyes, and turned his head, at first to one side, and then to the other, as if he were listening; he then opened his eyes, and said, disappointingly, "I cannot justly remember it." After a pause, his face brightened up a little, and, tapping the side of his head with the tip of the middle finger of his right hand, he remarked, laughingly, "I have it here quite well, if it would only come out," and immediately afterwards he became very animated, and brought down his hand on the knob of the bell, ringing it violently, while, at the same time, he shouted out, "Bell! There, I have got it." He repeats most words very promptly when uttered in his hearing. He even succeeded in pronouncing "Constantinople" fairly well, but stuck in the middle of "hippopotamus." The correct spelling of his surname not being known to us, I wrote on a piece of paper, "James Prophet," and asked him if that were correct. "Yes, I think that's it," he said dubiously. "Do you not think that this is right?" I asked, and wrote down "Profit." "No, sir," he replied hesitatingly, "I think that's it," pointing with his finger to the first name, but he then pointed to the letter "f" in the second name, and said, "There are two there." I now wrote down "Prophit," and he said promptly, "No, sir;" and again pointing to the first, he said, "I think that's it." I now spelt the name "Proffit," when he immediately called out, "That's it," and afterwards adhered to this assertion in a very positive manner. On being asked to read the heading of his prescription book, he surprised us by saying, "Manchester Royal," but he failed to make out "Infirmary." I now asked him to spell the word, but he was incapable of telling a single letter, and could not be got to hazard a guess. I pointed to the first letter, and asked him if it were "A." "No, sir," he replied. "Is it 'B'?" I asked. "No, sir," he said, laughing. "Is it 'I'?" was now inquired. "Yes, sir; that's it," he said, with emphatic assurance. It was the same with all the other remaining letters. He at once knew when they were correctly named in his hearing, and could immediately repeat them, but he was quite unable to name them correctly without prompting. He could not read a single word from a newspaper, or even recognise a single letter. On being asked to tell the time by a watch he at first failed, but after a time he drew, as he did on a previous occasion, his index finger round the dial, and counted from one up until he came to twelve, the hour hand pointing between eleven and twelve. After looking for some time at the hour hand, he drew his finger back to the minute hand, which was pointing between seven and eight, and after another considerable interval, he burst out with "Twenty minutes to twelve; that's it."

In the next case which I shall report word-blindness is accompanied by word-deafness. The patient is also suffering from bilateral hemianopsia.

Case 4.—Joseph Lander, aged 51 years, entered the Manchester Royal Infirmary under Dr. Ross, on August 20, 1884.

History.—The patient is a working engineer, and has been, according to his wife's statement, always a strong and healthy man. He has been married eighteen years, and has three children, all of whom are healthy. There is no evidence of syphilis. The patient has indulged freely in beer drinking, and has been a good deal exposed to cold and wet; but he has always had a comfortable home and been well fed. Three weeks before he came to the Infirmary he was from home and slept at Sheffield. He went to bed in his usual health, but in the morning he was unable to speak. All he could say was "home," "home," and he wanted to go to the station in order to get back to Manchester. He was, however, detained until his wife arrived, and she immediately got him to the train and brought him back home. For the first few weeks he was under the care of Dr. Scott, and during that time he could speak but little, and understood very imperfectly what was said to him. The day after he came home he wanted some article of diet. He kept repeating, "I want some of them there," and at the same time he put his hands together, and moved them as if he were breaking a biscuit in two. His wife presented to him every kind of diet she could think of, and even sent out for oysters and various other likely articles, but it was only at the end of five or six hours experimenting that she, aided by her youngest son, hit upon the diet he wanted, which consisted of boiled bacon and beans.

Present Condition.—The patient is a strong muscular man, and with the exception of the usual signs of arterial degeneration he is free from any organic disease. So far as can be judged, the patient is the subject of a right-sided bilateral homonymous hemianopsia; but it is so difficult to make him understand the necessary directions, and he is so confused in his replies, that it is impossible to attain to any certainty on this point. When asked to put out his tongue he instantly obeys, but does not comply with any other request unless the spoken demand is accompanied by expressive pantomime. He is incapable of naming correctly almost any object presented to him, or any of his bodily organs. His fingers he calls the first, the second, the third, and the fourth. Shown a bunch of keys, he took a hold of one between the finger and thumb of the right hand, and imitating the unlocking of a door he said, "It's a wheel." Not being satisfied with this name, he made several other attempts to re-name it, and as every fresh effort ended in his calling it "a wheel," he at last gave up the attempt, and said, "That's the way I am; I know it quite well, but I cannot say it." When asked to count numbers in succession beginning with one, he always says, "The first, the second, the third," &c., and by no persuasion can he be made to say "one, two, three," &c. He is wholly incapable of breaking in upon the series. When asked to say "three," for instance, he begins by "the first." Shown a shilling, he called it "the first;" then correcting himself, he named it "Saturday." Shown half-a-crown, he called it "two and two pence;" but when the shilling was again presented to him, he named it correctly. The half-crown being again shown him, he called it "two and two pence halfpenny;" but evidently recognising that he had made a mistake, he said, "That's what bothers me; I cannot say it." Shown a pencil, he said, "It's a pu——; it's a punt—; no, that's what bothers me," and he gave up the attempt. Asked the day of the week (Wednesday), he said it was "Waterday." He cannot read a word or recognise a single letter or number; but as he gets very irritable when pressed to read, it is possible that his incapacity may not be so great as it seems. He cannot be induced to attempt to write his name; and when asked, he protests loudly that he can't. The patient was very unsettled in the Infirmary, and was discharged on September 8th, 1884.

January 16, 1886.—Joseph Lander appeared occasionally at the Infirmary as an out-patient for some time after his discharge as an in-patient. About the beginning of 1885 his wife told me that he was "fretting at not getting to his work," and at her urgent request I wrote a note to his employers to say that I thought it likely that he would be found on trial to be quite capable of undertaking his previous work of taking charge of a stationary engine. He appears to have been an old and valued servant, and was accordingly re-instated in charge of his engine, and he has continued to perform his duties, without assistance, from that time until this date. Since he entered upon his old employment I have seen little of him until this evening, when he came for examination, in obedience to a post-card I sent him a day or two ago. In general health he is much the same as he was when he came first under observation, but looks considerably older now than he did then. An examination of his fields of vision shows that he is suffering from a complete right-sided bilateral homonymous hemianopsia, the blind being separated from the sensitive parts of each retina by a sharply-defined vertical line. His speech has improved considerably in several respects since last report. He can now name correctly many common objects, and several of his own bodily organs, but his vocabulary of names is still very limited. His nose, head, eyes, mouth he named correctly and promptly. Asked the name of his finger, he said, "That's my first;" but, on being pressed still further for a name, he replied, "That's the way I am ; I've forgotten it again, I know it sometimes quite well." On the name being uttered in his hearing, he recognised its correctness and repeated it. Asked the name of his ear, he replied by the usual formula of failure, "That's the way I am, I have forgotten it." "Is it your nose?" I asked. "This should be my nose," he replied archly, while at the same time pointing to the proper organ. Catching a hold of his ear, I said, "This is your ear. Say my ear." In reply, he said, "My nose, no ; that's not it," and after many efforts he was still unable to repeat the correct word. I then asked him to say the letter E, but in this also he failed. I now got him to repeat, after me, the alphabet, beginning with A, and when he got to E I stopped him, and asked him to say ear, but it was only after several repetitions of this process that he at last succeeded in getting the word. My first effort to get him to repeat a word after me was to ask him to say "Constantinople." "I could not say it," he replied, "if you were to give it me." On being pressed, he made one or two attempts, but never succeeded in getting the first syllable. Some short, common words, like ink, for example, he repeated after me without hesitation. His capacity to understand questions varies a good deal. On interrogating him, for example, about his vision, he gave a graphic account—partly in words and partly by acting—of how his visual defect caused him to come into collision with people on the street. "Little children," he said, "get right under me before I see them ; I am frightened of going over them." A few minutes before this he was bemoaning his condition, and said, "If you were to ax me my own name just now I could not tell you." I asked him if he could write it, and, with some reluctance, he took the pen and wrote—

When he came to the last letter, which was evidently intended for the first letter of his surname, he stopped, and, after several efforts to continue, he said, "No, I can't manage it." He now made a fresh attempt, and wrote—

but could go no further, and said "T'other part of the name I can't manage at all." During these attempts at writing his name he got very flushed in the face, and for some time afterwards he was very confused. I asked him, "Where do you live?" "What, what?" he asked, "my father?" His wife then asked him the same question in the vernacular, but he still kept asking in a confused manner, "What, my sister? my grandmother?" I then mentioned Ardwick, the locality in which he lived. "Oh, Ardwick?" he said, "Oh yes, I live in Ardwick," but he could not be got to name the particular street, even after it was told him. I now wrote—

Joseph Lander

plainly on a piece of paper, and asked him to copy it, and after considerable hesitation and trouble he wrote—

J. oudpoh   Lander

Finding that the copy had assisted him in writing his name, I asked him to copy the word "contents," which was printed in moderately-large capitals in a newspaper lying on the table. He was, however, unable to copy a single letter from the printed page, but when I wrote the word—

Contents

on a piece of paper, he wrote below it the following—

Carebrobra

which is a fair attempt, although it would not be easily deciphered if not for the knowledge that it was an endeavour to copy the word "contents." I now drew a square on paper, and he immediately made a very accurate copy of it, but was unable to name the figure, and added, "I know it quite well; I used to do a lot of this at our work," meaning that drawing was a necessary accomplishment for an engineer. I then drew roughly the outline of a man's face, and he immediately took the pen and quickly drew another, and laughingly remarked, "I have put plenty of nose on."

The patient is quite unable to read a single word. He can, however, recognise some letters of the alphabet, but not others. My eye lighting upon the word "whisky," printed in moderately-sized capitals on the cover of the *Graphic*, a word doubtless at one time very familiar to him, I asked him to spell it. He spelt it, "H, h, i, s, h, e." His mistakes having been corrected, he at last succeeded, after many trials, in spelling it correctly, but very slowly. He voluntarily continued the exercise, and at each repetition he spelt the word quicker, and with greater ease, and after several repetitions the letters followed each other as quickly as in the usual spelling of healthy people, when at last his countenance lit up with a smile, and he pronounced "whisky."

We have already seen that in Case 3, and still more markedly in Case 4, the word-blindness was accompanied by a considerable degree of word-deafness, and we shall now proceed to describe a case of more or less pure word-deafness, or, at any rate, one in which the inability of comprehending spoken speech was not accompanied by any marked difficulty in understanding written language. It may, however, be observed that, inasmuch as language is first developed in connection with the sense of hearing, injury of the auditory mechanism may be expected to produce a much more profound disorganisation of the faculty of speech than does injury of the visual mechanism. An imperfectly educated man, for example, scarcely comprehends written speech unless he reads aloud, and in him it is manifest that loss of the faculty of appreciating spoken speech would necessarily carry with it loss, or, at least, great diminution, of the power of understanding a printed page.[1] It has already been remarked that an injury of the impressive mechanism of speech will throw into disorder the expressive mechanism, and it is in cases of word-deafness that we meet with the most marked examples of the condition which Kussmaul has called *paraphasia*. In this condition the patient applies wrong names to objects, using a word kindred in its meaning with the one intended, as "worm powder" for "cough medicine;" or in its sound, as

[1] See Watteville, Dr. A. de. "Note sur la cécité verbale."—*Le Progrès Médical*, March 21, 1885

"parasol" for "castor oil," or "butter" for "doctor." A still more profound disorder of the impressive faculty of speech is reached in those cases, which are by no means rare, in which the patient is unable to name his own bodily organs, or any of the common objects by which he is surrounded. Inability to name objects has been called by Kussmaul the *aphasia of recollection.* We accept the term simply as expressive of certain clinical facts, although it must be admitted that it carries with it somewhat undesirable theoretical implications. With these preliminary remarks we shall proceed to report the following example of *word-deafness,* which is accompanied by a profound degree of the aphasia of recollection.

Case 5.—James Lee, aged 57 years, is an out-patient under me at the Manchester Royal Infirmary at the date of this report, December 16th, 1885. He has, however, been more or less under my observation for the last four or five years, having been two or three times an in-patient, and many times an out-patient, of the Royal Infirmary, as well as a patient several times in the Barnes Convalescent Hospital at Cheadle.

Previous History.—The patient was born in Rugby, and was a gas-fitter by trade, but for many years he has been a messenger at the Manchester Bankruptcy Court. He was always an unsteady man, and his employment, in serving writs for the last few years, favoured his drinking propensities, and he was a well-known character in all the beer-houses in Manchester and neighbourhood. He has a grown-up family, all of whom are healthy, and there is nothing in his history pointing to syphilis. When he came for the first time under observation, four or five years ago, his wife stated that he had a fit several months previously, and from that time he had been unable to speak correctly. She asserted that he was not completely unconscious or convulsed during the attack, and that he never had any signs of paralysis. On admission to the Infirmary, he was found to have an earthy complexion, his temporal arteries were dilated and tortuous, the radials were hard and knotty, and the second sound of the heart at the base was highly accentuated. The urine was free from albumen or sugar. The grasp of either hand was good, and there was no exaggeration of the tendon-reactions at either knee or wrist, or any other evidence that the patient had suffered from paralysis. He sees imperfectly with the left eye, but we gather from him that this defect is of old date. The media of this eye are transparent, and, with the exception of slight pallor of the disc, no changes are discovered in the back of the eye on ophthalmoscopic examination. The right eye is the subject of a high degree of presbyopia, but there is no restriction of the field of vision, and the optic disc is healthy. On applying a piece of camphor to the left nostril, the right being closed, he gives no evidence that he perceives the smell, but when he sniffs it with the right nostril he immediately draws his head back, and, pointing with the index finger to the right nostril, he says, "This is all right, but," he continues, pointing now to the left nostril, "with this one I can't at all." This test has been frequently applied during the many years the patient has been under observation, and always with the result just described. His sense of hearing for ordinary noises is very acute. He hears the ticking of a watch when it is a foot or more from either ear, and the slightest knocking at a door, or other noise at his back, causes him to turn round. No other sensory disorders have been discovered. The patient, on being interrogated with regard to his occupation, replied, "When gentlemen got into trouble, I brought an action against them, and then they were all right;" and in order to make still plainer the functions he exercised in connection with the Bankruptcy Court, he generally produced out of his pocket a writ or some such legal document.

On being asked the nature of his complaint, he replied, "I am all right, if I could only speak it; there is nothing the matter with me. I am quite right, but I cannot speak it at all." He was now handed a bunch of keys, and asked to name one of them. He held one between his thumb and index finger, and said, "It is a public-house." "That is not a public-house," I said. "I know it quite well;" he replied, "I have seen it thousands of times," and, trying again to name it, he continued, "It is a—it is a public-house. Pooh! I know it quite well;" and, by way of further explanation, he added, "When you go to bed at night, you do this," turning the key from left to right, as if he were locking a door, "and when you get up in the morning, and want a drop of beer, you do this," turning the key from right to left, as if unlocking. A latch key was now placed in his hand, and he was asked to name it. He immediately assumed a very comical expression, and said, "I know it quite well. When gentlemen, you know," he now looked still more comical and knowing, "are out, and when they come in, they do this," imitating the unlocking of a door, "and then they are all right." He was now asked to name a watch key, which was placed in his hand, and he instantly replied, "It is for your public-house, there," pointing to my watch pocket, "you do this," imitating the movement of winding a watch. Asked to name the index finger of his right hand, he said, "That is the first one." "The first what?" I demanded; to which he replied, "The first public-house." The middle finger was named "the second," the ring finger "the third," and the little finger "the fourth public-house." He managed to name coins fairly correctly, although not until after several futile attempts. Half-a-crown was often at first said to be "twopence halfpenny," and a two shilling piece, "twopence;" but when coppers were shown to him he generally named them correctly, and then sometimes, but not always, he worked up to a sixpence, a shilling, two shillings, and half-a-crown without committing a mistake. Every other object presented to him was "a public-house," or "a glass of beer." A very laughable illustration of the manner in which he used the word "public-house" on all occasions, was afforded in 1883. I brought him over, along with several other patients, to show to the members of the British Medical Association attending the Liverpool meeting. Dr. W. H. Broadbent, who took a great interest in his case, had just finished his examination of him, and we turned away to examine a case of rupture of the brachial plexus. James Lee advanced and tapped me on the shoulder, and pointing to the other patient said, with an air of great importance, "I know this gentleman, I know him quite well; he once lived in the next public-house to me." The other patient explained that eighteen months previously he occupied for some weeks the next bed to Lee in the Infirmary.

The patient is quite unable to repeat any name uttered in his hearing, and does not even seem to appreciate the difference between the right and the wrong name. When, for instance, he holds a key in his hand and is asked whether or not it is a horse, he gets confused and says hesitatingly, "Yes! Yes, it is a public-house," and when he is next asked if it is a key, he gives the same answer in much the same manner, without giving any evidence that he has recognised the absurdity of the first and the appropriateness of the second designation. He can count in succession numbers up to twenty or even higher when he starts with one, but he is unable to break in upon the succession. Asked to start with "five" for example, he begins with "one." When asked to put out his tongue in spoken language he promptly opens his mouth and complies with the request, but this is the only mandate he obeys unless the spoken request is accompanied by an expressive look or other sign. He is, however, very quick at interpreting pantomimic speech. Asked his age, he said, "I think I am twenty; no, that is not it, I think I am twen—. No, I can't speak it at all, I know it quite well." "Don't you think you are a hundred?" I asked. "No, I don't think I am so old as that," he replied, somewhat dubiously. After being asked his name and age several times, he came one day armed with a small pocket-book in which he had his own name and address, his age, my name, Rugby, which was his birthplace, and the names of several other places and events written down carefully. When now asked his name or age he immediately dived into his pocket and producing his book he pointed to the writing.

"See, that is my name. James Lee is my name ; I was born there—Rugby. That is my age ; I am 57 years of age." Asked now if he does not think he is a hundred, he says, "No, I am not so old ; that is my age," pointing to 57 in his book, but at the same time he does not appear to be in any way sensible of the ridiculousness of the question. When, however, the number 100 is presented to him in written characters and he is asked if that figure corresponds to his age, he immediately bursts out laughing and says, "Pooh! pooh! nothing of the kind." Pointing again to the 57 in his book he says, "That is my age." Although James Lee will not obey a simple request like "Give me your hand," he understands at other times apparently much more complicated statements and requests. His wife told me that he was very unruly at home, and that he often struck at her and his daughters. For his conduct to his family I gave him a severe reprimand, the nature of which he appeared to understand perfectly. Some time afterwards I asked him "Do you strike those at home yet?" He blushed deeply and said with a forced smile, "Yes, I give them a clout (slap) sometimes." Then getting very angry and losing his self-control he continued, "Yes! they don't do well to me, but I will kill some of them, I will," he continued with fearful energy and determination, "I will kill some of them yet." "If you kill any of them," I said, "you will be hanged for it." "What do I care?" he retorted ; "they can't do anything to me ; I don't care a bit, it is nothing to me, it will be done with then," evidently meaning that he did not value his life, and did not care how soon or in what manner it was ended. On now being asked to show his hand he got confused, and after a time put his tongue out. I then repeated the question, and both looked at and pointed to the hand, and he immediately held it out, showing conclusively that his want of compliance with the first request was simply because he did not understand its terms. He evidently understood the nature of my threat that unless he restrained his passions he would get hanged, while he was unable to comprehend the name of his hand.

James Lee can write in a beautiful round hand, which is singularly free from the tremor of age or from the signs of any other infirmity, and he can copy from a printed or written page without committing mistakes in spelling or punctuation. He is, however, unable to write a single word to dictation except his own name, which he always writes, when asked, with apparent pleasure and pride. When requested to add his address to his name, he produces his little pocket-book, and carefully copies the address. His efforts at spontaneous writing are somewhat ludicrous. A short time ago I sent him a post card to ask him to come to the Infirmary, as I wished to show him along with other cases of

c

aphasia to a clinical class, and by return of post I got the following post card :—

5 Pownall Street, Hulme Decr 12 1885

Dr Ross

Yas Yours truly Yours trust James Lee

The card was doubtless intended to intimate that I might rely upon his appearing at the appointed time. During the summer of this year (1885) James Lee was an in-patient in the Infirmary in the same ward with a very intelligent patient who became interested in his case, and was very attentive and kind to him. After a time Lee was sent to the Barnes Convalescent Hospital, and soon afterwards the other patient, who remained in the Infirmary, showed me the following letter, which he had received from him :—

Dear Sir I canot to bor me I all right for My exe for my furty to mes, I manghet to My "The Rugby" I hope too yours all Yours Truly James Lee

The recipient of the letter thought, and I have no doubt correctly, that Lee meant to thank him for his exceeding kindness to him. Along with the letter came an old newspaper entitled *The Rugby*, and the fact that these words were copied from the printed page explains how they came to be correctly spelt.

The patient reads fairly well, but it is difficult to determine how much of it he understands. When asked to read a paragraph of a newspaper, he reads on without much hesitation, but he seldom completes a sentence without committing a mistake or two, which completely mars the sense; but he goes on without trying to correct himself, and when he has finished, he cannot be got to give any indication that he has comprehended the nature of the subject. He cannot also be got to obey written requests. I have tried him with such a simple request as "Put your tongue out" written plainly on a slip of paper. He read the words at once, but put the paper down without giving any sign that he understood the nature of the request, and although he always gave prompt obedience to this mandate in spoken language, he could not be got to obey it when written. It is probable that on being requested to read before spectators, he regards the simple act of reading as an exercise to be performed without any ulterior aim, and that consequently he makes no effort to comprehend the meaning of the composition. There can, indeed, be little doubt that he does understand a good deal of what he reads in private, although it is possible that his comprehension of much of it is somewhat dim and obscure. During the public excitement caused by the revelations in court at the trial of the Phœnix Park murderers, at Dublin, he came to the Infirmary one morning and said to me, with fierce energy, "Those fellows over there—I would hang every one of them, every one of them I would hang." I have no doubt whatever that by "those fellows over there" he meant the Irish prisoners, and possibly all their sympathisers, and he must therefore have understood the nature of their crime from reading the newspapers, because his comprehension of it through spoken speech was wholly impossible.

When an in-patient of the Infirmary he is generally found with a newspaper near him, and he frequently scans its columns. One day I remarked to him, "I see you are reading your newspaper." "Yes," he replied ; "but I cannot speak it at all." "But you understand all that you read?" I said, enquiringly. "No, I can't," he rejoined ; "I can't speak it." He then, by way of correcting himself, pulled a very sorrowful countenance and said, "No, I can't do it ; I can't do it at all. I used to do it very well, but I can't do it now at all." From his manner, which long observation had often enabled me to interpret readily, it was evident that he meant to make a distinction between the "I can't speak it" and "I can't do it." By the last phrase I have no doubt he meant me to understand that he was not able to comprehend the meaning of what he read now as he once did. His comprehension of some written words was, however, made convincingly clear to me in another way. For some time now he has ceased to name everything presented to him "a public-house," but calls them "a glass of beer," or "a drop of beer" instead. One day I asked him the name of his index finger, and, as usual, he said, "That is the first one," and on being pressed for a further name he called it "the first glass of beer." I then wrote down "finger" on a piece of paper, and on reading it his face lit up with intelligence, and pointing to the index finger of his left hand, he said "That is this—that is my finger." After a short time I again asked him the name of his index finger, and he immediately cast his eyes about in search of the paper, and when his eye lit upon it, he said triumphantly, "That is my finger." The paper was now removed, and after another interval he was asked the name. He now, as before, cast his eyes about him in search of the paper, but on failing to find it, he looked in the distance as if he were in deep thought, and kept repeating at intervals, "It's my fe— fe— fe—", but on being again urged to name it, he looked down at his finger and gave the stereotyped answer, "It is the first glass of beer." It was found that he could readily identify the written names of key, book, pencil, &c., with their corresponding objects. The request to name his finger being a favourite test and frequently used, it was found after a time that, without being shown the written word, he made a strenuous effort to recall the name by repeating "f— f— f—," but this effort always ended in his calling it the "first glass of beer." When once, however, he saw the printed word he retained its use without refreshing his memory by sight during the rest of the time occupied by that day's examination. There can be little doubt, therefore, that the names of common objects could have been reorganised in this man by the systematic use of printed words.

January 9th, 1886.—James Lee came to the Infirmary this morning as an out-patient, and his behaviour forms a striking commentary on the remark just made on the possibility of reorganising in him the names of common objects. On testing him in the usual way, by asking him to name his index finger, he replied by calling it his "hand," instead of "the first glass of beer." I asked him to name his hat, which at the time he held in his hand. "That is," he said with some hesitation, "that is my head," and at the same time he lifted his hat and put it upon his head. A son-in-law of his, who accompanied him, said, "You should say hat; you knew the name quite well last night." I then said, pointing to the hat, which he now again held in his hand, "That is your hat, and not your head." "Yes," he said, "it is my—— it is my head," again raising the hat and putting it on his head. The son-in-law again interposed, and said to me, "If you will write down the name, he will know it at once. I tried him for a long time last night, and by writing down the words for him he got to know hat, head, hand, and various other names quite well; but he has forgotten them now." The son-in-law also stated voluntarily, "His judgment is very good. He is, indeed, a man of very excellent judgment; there is nothing the matter with him except in his talk." This shows that a man of ordinary intelligence, and without any special experience or training, may recognise the difference between the mental condition of this patient and that of an insane person.

January 27.—The patient was once more tested to see if he could be got to obey written requests. I wrote down on a piece of paper, "Put out your tongue." He read it at once, but put down the paper as usual without complying with the request. I kept pointing to the paper and urging him to do what was wanted. After a considerable time his face lit up with an intelligent smile, and he obeyed. He was now presented in succession with the written requests, "Show me your hand" and "Where is your head?" and to these he promptly replied by holding out his hand and pointing to his head; and on being asked in writing, "Have you got any keys?" he put his hand in his pocket and produced one. His ready compliance with these requests shows pretty conclusively that, as I have already remarked, his failure hitherto to obey written commands was owing to the fact that he regarded the words presented to him simply as an exercise in reading.

In the following case of word-deafness the incapacity of the patient to express his thoughts in spoken language is much greater than in the case just reported.

Case 6.—John Morris, a man between thirty and forty years of age, entered the Manchester Royal Infirmary on July 26th, 1886.

The patient is unable to give any account of himself, and all that is known of him is that some months ago he had "a stroke," and that for some weeks afterwards he suffered from some degree of paralysis of the right side of the body, and loss of speech, being unable either to speak himself or to understand what was said to him. He gradually recovered the use of his limbs, and his mouth, which was at first drawn to the left, soon became straight.

Present Condition.—The patient is, on the whole, a healthy-looking man, but his pulse is found to have the usual typical characteristics of Corrigan's pulse, and his forehead, on being scratched, presents well-marked capillary pulsation. On physical examination of the chest, the apex of the heart is found to be displaced downwards and to the left, the præcordial dulness is enlarged, and a double murmur is heard in the aortic area, the diastolic murmur being heard at the top of the sternum but not in the carotids. The patient presents, in short, all the symptoms of a free aortic regurgitation, but in other respects his general health is fairly satisfactory. When a piece of camphor is applied to each nostril in succession he indicates plainly by pantomime that he appreciates the smell with his right but not with his left nostril. The patient puts out his tongue when asked to do so, but he does not obey any other spoken request. To all enquiries about his health he replies in such phrases as "Yes, thank you;" "John, if you please, ma'am;" "All right, thank you;" but his reply never seems to have any kind of relation to the question asked him. On being asked, for example, to name the index finger of his right hand, he says, "John, if you please, ma'am." His attempts at reading are invariably the same. He points with the tip of his right index finger to each word in a line in succession, and repeats, "John nissus, John nissus, John nissus, bree." This jargon is intoned as a kind of chant, and the last word is dwelt upon with an air of much apparent satisfaction, and prolonged to an unusual degree. On uttering "bree" the patient stops, hands back the book or newspaper from which he has been reading, and says, "Thank you, if you please, ma'am," and if urged to continue his reading he repeats the same jargon, beginning as before with "John" and ending with "bree." At times he manifests signs of irritation and disappointment at not being able to read. On these occasions he stops short at "John" or "nissus," utters "Tut! tut!" knits his brow, and assumes a lachrymose expression. He at the same time gives a significant jerk of his head to the left, and points with the index finger of his right hand to his tongue or the roof of his mouth, as if he wished us to understand that the cause of his misfortune was to be found somewhere in the cavity of his mouth. But although the patient is unable to read aloud, he immediately obeys written requests. When such words as "finger," "hand," "head," "ear," "table," &c., are presented to him he immediately points to the right object, and gives prompt obedience to written requests like "shut the door." He is unable to write any word, either spontaneously or to dictation, except his own name, which he writes in very legible characters. He can, however, copy a printed word in printed but not in written characters, and a written word in a current hand but not in printed characters. He is also able to add up a column of figures, and to write the result in figures correctly.

Of the few sounds which the patient uttered on attempting to read, the origin of John as being his own name is apparent. The *nissus* was probably suggested by the hissing sound of the "s" in Morris, but it is hardly possible to give any account of the sound *bree*.

In the following case, reported by Giraudeau,[1] the patient could communicate her ideas in writing, but it is not mentioned how far she had the power of naming objects.

Case 7.—"Bouquinet Marie, aged forty years, laundress, entered L'Hôpital Saint Antoine, under the care of Professor Hayem, on February 22nd, 1882. The patient had never suffered from any serious disease previous to her present illness, and nothing in her history pointed to alcoholic excess or to syphilis. She never had been regular, and had ceased menstruating for six months. For three months she had suffered from a constant headache, which was diffused over both sides of the head, and was subject to a nocturnal exacerbation of such severity as to have prevented sleep. In the week previous to her admission, the headache was at times so violent as to cause the patient to cry out with pain. The patient had never suffered from vomiting, loss of consciousness, or epileptiform attacks. About a month previous to her admission, the pains became so severe as to compel the patient to give up work; and at the same time it was observed that she was no longer able to understand what was said to her, and did not answer when spoken to, but she never committed any unreasonable action. She passed her time at home bemoaning her condition, and only occasionally went out. These facts were obtained from the person who accompanied her to the hospital, the patient herself being unable to relate her history.

"*Present Condition.*—The patient is very stout, there is no fever, the right pupil is slightly dilated, and the presence of violent headache is attested by the hand being frequently placed upon the head. Asked her name, she raises her head but gives no reply. Asked a second time, she demands, "What do you say?" and on the question being put a third time she says, "I do not understand you;" but when pressed a fourth time for an answer, she generally gives her correct name, "Bouquinet Marie." On being asked, "How long is it since your illness began?" she manifests the same difficulty in understanding the question, but after a time she replies, "For three months." If she is now asked her address, she replies, "Probably for three and a half months." Interrogated with regard to her occupation, she presents the prescription of the physician who had treated her in the town, and adds, "A white powder" (sulphate of quinine). The questions addressed to her were varied from time to time, but her replies were always more or less similar to those reported above. After getting her with much difficulty to comprehend the first question addressed to her by repeating it frequently, she often answers

[1] Giraudeau (C.). "Note sur un cas de surdité cérébrale (surdité psychique) par lésion des deux premières circonvolutions temporo-sphénoïdales gauches."—*Revue de Médecine*, Tome 11. 1882, p. 446.

correctly, but she then follows the idea started, and her subsequent replies have no relation whatever to the questions addressed to her. At times, however, it is impossible to get her to comprehend any idea, and to all questions she replies, "What are you saying? I do not understand. Cure me." Her sense of hearing is at the same time intact; there is no discharge from the ear. She hears the ticking of a watch, and turns her head when a slight noise is made near her. Vision is good in both eyes. There is no word-blindness, and it is important to observe that she can easily read the headings of the bed ticket; and to questions addressed to her in writing she replies, after a short pause for consideration, either in a lively voice or in writing. It is in this way that a knowledge of the fact that she had never been regular, and that she had ceased to menstruate for six months, had been obtained. Tactile sensibility is preserved, while the senses of taste and smell are normal. Motor power is intact on both sides, and the tendon reactions are normal."

It is unnecessary to describe the details of this case further, inasmuch as there was no serious alteration in the state of her speech up to the time of her death, which took place from coma about a week after her admission. At the autopsy a sarcomatous tumour was found in the left hemisphere, occupying the position of the first and second temporosphenoidal convolutions. The patient, whose case has been reported by M. Giraudeau, appears to have had the power of conveying her ideas in writing in a much greater degree than had James Lee; but the report is not sufficiently definite to enable us to say whether she could name objects at sight. Her inability to give a connected history of her illness in spoken speech indicates that the impressive faculty of speech was profoundly affected, and as the power to name objects is probably the most vulnerable part of speech it was probably lost in this case.

Although reported cases are not always so definite as could be desired it appears to me that an examination of them will show that word-deafness necessarily carries with it the presence of the condition in which the patient applies wrong names to objects—*paraphasia*, or in which he is totally unable to give any name to an object, these two conditions being only different degrees of one and the same disability. The converse of this proportion is, however, not true. It is possible for a person to labour under at least a considerable degree of inability to name objects, and yet to be free from either word-deafness or word-blindness, as the following case testifies:—

Case 8.—R. B., a man about 45 years of age, came as an out-patient to the Infirmary, on January 30th, 1866. He said that he had lost his memory, and was unable to give a connected account of himself. We gathered from him that at Christmas he suffered from some cerebral attack which rendered him unconscious for two or three weeks, and that "ice," a word which he frequently repeated, was applied to his head. He could not remember the name of the doctor who attended him, and was even unable to recall the name of his wife or of any of his children; but he told us his own name and the district, but not the street, in which he lived. He told us that he had been married many years, and that he had a large family; but he was unable to give us any further trustworthy information with regard to his history. The tendon reactions were normal, and there were no signs of motor paralysis, while the cutaneous sensibility and all the special senses were found normal. The urine was free from albumen, the cardiac sounds were normal, and the usual signs of extensive arterial degeneration were absent. The patient gave prompt obedience to any spoken request, so that he was free from word-deafness. He was also able to name his bodily organs, although not without considerable hesitation. He likewise named correctly, but always after a long pause for consideration, several objects presented to him; but he was quite unable to name others. Given a bunch of keys he took one of them between his finger and thumb, and moved it as if he were locking a door; he then looked in the distance as if intently listening, but not being able to recall the name he again looked at the key, and still not succeeding, he looked in the distance a second time, and after another long pause he burst out with "key," his previously perplexed face lighting up with pleasure. A watch key was now placed in his hand, and he immediately rotated it as if he were winding a watch, and went through the process of looking alternately in the distance and at the key, but all to no purpose. After a time he pointed to my watch pocket and said, "It is for that;" and he again rotated the key, as if he were winding the watch. I now produced the watch, and without much hesitation he called out "watch." He also read off the time correctly, but was still unable to name the watch key. I purposely suggested wrong names for the key; he promptly rejected them, but joyfully assented when the correct name was given, and he immediately repeated the name. After a short pause the watch key was placed in his hand a second time, and again he was unable to name it. He was now asked if he knew the name of the building in which he was standing, but he was unable to tell. On being told that it was the Infirmary he readily assented; but on being asked almost immediately afterwards to repeat the name of the building he was unable to do so. He was now requested to read, and when the heading of his prescription book was shown to him, he read without the slightest hesitation, "Manchester Royal Infirmary." The book was now withdrawn, and he was asked to repeat what he had read. He looked in the distance, as he always did when endeavouring to recall a name, but he was quite unable to recall a single word of the "Manchester Royal Infirmary," which he had just read. At the same time he gave prompt obedience to written requests, so that he was not suffering from word-blindness. The age of the patient, his freedom from cardiac or renal disease, and the absence of the usual signs of a cerebral tumour led me to believe that his symptoms might possibly be caused by syphilitic endarteritis, and accordingly he was ordered 10grs. of iodide of potassium three times a day.

February 6.—The patient came to-day for his medicine, and his disorder of speech was found to have undergone much improvement. He now told us that he had been married over 20 years, and that his wife had 14 children, but that only seven of them were living. Of the remaining seven some were still-born, and others died from convulsions in early infancy. The result of the treatment not only confirmed the diagnosis of a syphilitic taint, but also showed that the disorder of speech was most probably caused by a profound anæmia of the cortical mechanism in the absence of any permanent destruction of tissue. The case which has just been reported is a good example of verbal amnesia, or what Kussmaul calls the *aphasia of recollection*, and with it we must bring to a close our remarks on the clinical aspects of sensory aphasia, and proceed to describe briefly those profound disturbances in which the sensory and motor mechanisms of speech are simultaneously affected.

3.—Combined Sensory and Motor Aphasia.

In many cases of what turns out after a time to be a pure motor aphasia, the disorder of the expressive faculty of speech is accompanied during the early stages of the affection by a manifest disturbance of the apperceptive faculty. In the case of William Davies (Case 1), for example, his wife stated that for the first few weeks after his attack he was not only unable to speak himself, but that he was also incapable of understanding the spoken language of others, showing that he was the subject of word-deafness as well as of motor aphasia. And when he came under my observation, some months after the attack which ushered in the aphasia, he could not be induced to obey written commands, so that most probably he was then suffering from some degree of word-blindness. This patient was examined by me a few days ago, when I found that he gave prompt obedience to such simple written requests as "Put your tongue out" and "Show me your hand," proving that he has recovered, to some extent, from his word-blindness, and that his case is now to be regarded as an example of pure motor aphasia. Some months ago I had an opportunity of observing, day by day, the gradual recovery of an aphasic patient from the disorder of the apperceptive faculty of speech, while that of the expressive faculty persisted unchanged for months. The patient was a boy about ten years of age, who was under the care of my colleague, Dr. Morgan. The boy was attacked with a slight apopleptic seizure, and became suddenly aphasic and paralysed on the right half of the body. A soft systolic blowing murmur was audible over the apex of the heart, and consequently the attack was probably caused by embolism of the left Sylvian artery. For the first few weeks the patient was not only unable to utter a single word, but he also failed to obey the simplest request, such as "Put your tongue out," provided the spoken request was unaccompanied by any expressive gesture. When the observer, however, put his own tongue out, and made signs to the patient to follow his example, prompt obedience was secured, and in a very few days he acquired the power of appreciating and obeying the spoken request without the aid of gesture. And after a process of training, which extended over three or four weeks, the patient came to recognise and to point out his own bodily organs, and the various objects by which he was surrounded, when their names were uttered in his hearing. The patient had been to school and been taught to read, but for the first few weeks he took no interest in one of his school books which was placed on his bed. But after a time he began to turn over the leaves, and from the frequency with which he amused himself with looking over its pages there can be little doubt that he understood the text. In this case the gradual disappearance of the word-deafness and word-blindness could be traced day by day. The motor aphasia remained unchanged for many weeks. It is worthy of note that the paralysed half of the body regained motor power *pari passu* with the disappearance of the sensory aphasia, and six weeks from the commencement of the attack scarcely a trace of paralysis could be discovered. It is equally important to observe that, at least in many of the cases in which a motor aphasia occurs without being accompanied by hemiplegia, the disorder of speech is from the first limited to the expressive faculty, and the usual evidences of sensory aphasia are altogether absent. The fact that motor aphasia with hemiplegia is sometimes, at least, accompanied by a sensory aphasia is proved by the statements of patients who have made a complete recovery. A case of this kind, reported by Charcot, is of so much importance that we do not hesitate to give it in detail.

Case 9.—" Louise Jeanniot, aged 49 years, entered the Infirmary of the Salpêtrière, under the care of Professor Charcot, on August 26th, 1883. On the previous evening, soon after retiring to bed, she suddenly lost consciousness. After a time, the duration of which she does not know accurately, she regained consciousness, and then suffered from a pain in the abdomen about the umbilical region, which was so severe that nothing was able to allay it, and which evoked from her piercing cries the whole of the night. The patient is laid upon her back with tympanitic abdomen, and suffering from intense dyspnœa, but she yet gives clear and accurate replies to all questions addressed to her, both with regard to her present condition and her previous diseases, &c. The urine is suppressed. The patient has had many involuntary evacuations from the bowels during the night, while a string of blood flows constantly from the anus. The inferior extremities and the whole of the abdominal walls are livid and cold. The pulse is small, thready, and uncountable at the wrist, and cannot be felt in the crural arteries. It was concluded that the abdominal aorta was obliterated. Jeanniot is the subject of a right-sided hemiplegia, with contracture, which implicates the side of the face as well as the limbs. Six years previously to her admission she was attacked during the night, and on waking in the morning she found herself paralysed and deprived of speech. For two years she was only able to pronounce 'yes' or 'no,' but she gradually regained her speech. The patient is unable to tell the order in which the words became restored to her. At present her speech is as correct and intelligible as that of a person who had never been deprived of it. She writes a little with the left hand, but not in the form of mirror writing. This morning she endeavoured, but without success, to write her name with the left hand.

"Jeanniot has always been able to read, and she read very well in our presence, both from a printed and a written page. She understands perfectly all our questions, and her replies do not indicate any incorrectness. It was not, however, always so. At the commencement of her hemiplegia, and during a period of which she is not able to determine the duration, but which was at the same time much shorter than the duration of her motor aphasia, Jeanniot only occasionally understood the words addressed to her, and therefore it was necessary that they should repeat things to her many times, and that she should fix her look carefully upon the person who addressed her. Speech was only for her an indistinct noise, similar to that of a conversation in a crowd. She at the same time heard the noise of a step in her room, of articles of furniture and utensils on being moved, and also of the door on being closed or opened. At that time she lived near the Square Parmentier, where the military band played regularly during the summer, and she went there with great pleasure to listen to it. They led her there several times when first she was able to go out. When the music played she heard distinctly that a new sound became super-added to those which she had before perceived, but it was only a noise which had no musical character. She died at three o'clock in the afternoon." At the autopsy, besides obliteration of the abdominal aorta, an extensive softening was found in the left hemisphere of the brain. This softening had destroyed the head and foot of the third frontal convolution, the whole of the Island of Reil, the whole of the first temporosphenoidal convolution, and a small portion of the inferior parietal lobule. On horizontal section of the left hemisphere, the claustrum, the anterior portion of the lenticular nucleus, and the anterior two-thirds of the internal capsule, were found to have been destroyed. The right hemisphere was normal.

In the case just reported, and in other similar cases, the motor disorder of speech persisted for a long time after recovery had taken place from the sensory disability, but in other cases of combined sensory and motor aphasia it appears to me that the disorder of the impressive faculty of speech is even more profound than that of the expressive faculty, although recovery cannot take place from the latter disability so long as the former remains. In the cases we are now about to consider the patient is not reduced to the condition of only being able to articulate one or two simple words like "yes" or "no"; he can give utterance to a large number of syllabic sounds, but he is unable to connect them into any intelligible word, and consequently his vocalisations are mere gibberish. This kind of aphasia, first described by Dr. Broadbent, may therefore be called *gibberish aphasia;* it is always associated with a profound disorder of the apperceptive faculty of speech, and with a persistent hemiplegia and contractures. The following is a good example of this kind of aphasia.

Case 10.—John Handforth, aged 52 years, appeared for the first time in the out-patients' department of the Manchester Royal Infirmary in May, 1885.

History.—The following brief account of his previous history was obtained from his wife. They have been married for upwards of twenty years, and have five children, all of whom are now living and healthy. The patient is a joiner by trade, but for some years has been an overlooker for a firm of contractors. He has enjoyed good health, but has always partaken somewhat freely of alcohol, and been much exposed to wet and cold. On November 12th, 1884, he went to bed apparently in good health, and got up on the following morning at the usual time without making any complaint. Soon after he got on the floor, his wife noticed that his right arm was drawn backwards behind his body and upwards towards the shoulder blade, and on looking at him more narrowly she found that his face was drawn, and that he was unable to speak. She got him to bed as soon as possible, and she soon afterwards observed that the right side of the body was completely paralysed. At first he was unable both to understand what was said to him and to express his thoughts, but she does not think that he was completely unconscious at any time, except, probably, for a very short period during the onset of the attack. He lay in this condition for many weeks without any very material improvement taking place. A few weeks from the commencement he had a severe attack of general convulsions, which lasted from five to ten minutes. During the attack the patient was profoundly unconscious, and bit his tongue, and the wife does not think that the one side of the body was affected by the spasms to a greater degree than the other side. He had another attack of the same kind three months after the onset of the disease. After the last attack he gradually improved, so that he was able to get up and walk about a little, but the right leg dragged, and the right hand remained stiff and helpless, though there did not appear to be any drawing of the face. His speech, however, has not undergone much improvement, and the little amendment which has taken place consists in his understanding somewhat better than he did at the first what is said to him.

Present Condition.—On his first appearance as an out-patient he was very excited, and by rapidly pointing, in a very animated manner, first to his paralysed hand, and then to his tongue, he made a strenuous endeavour to make me understand the nature of his disabilities, while, in the interval between these gestures, he gave utterance to volleys of unintelligible jargon. His general health appears to be good. His face has a mottled appearance, and his cheeks and nose are covered by dilated capillaries. The temporal arteries are distended and tortuous, and the radials feel knotty and resisting. The heart is found to be of nearly normal size on percussion, and no murmurs are heard on auscultation, but the second sound at the base is much accentuated, and of a ringing, metallic quality. The urine is free from sugar or albumen, and no changes are discovered in the back of the eye on ophthalmoscopic examination. So far as can be ascertained, there is no special defect of vision, and no loss of smell. The right leg drags, and the patellar tendon-reaction in it is exaggerated, while ankle-clonus can be elicited. He can move the right arm fairly well at the shoulder joint, but the forearm is held more or less rigidly at right angles to the arm ; the thumb is flexed into the palm, and the fingers flexed over the thumb. The tendon reactions are exaggerated at the elbow joint and wrist ; the hand is blue and cold, and the patient is unable to use it. Whilst the patient was being examined, he was seized with a severe general convulsion, during which he was profoundly unconscious; the tongue was protruded and bitten, and the spasms were distributed over the body generally, the right side being chiefly affected by tonic, and the left by clonic, spasms. The attack lasted about five minutes, and after its cessation the patient remained in an unconscious condition with stertorous breathing. He was consequently made an in-patient, but, on regaining consciousness, he showed signs of impatience at being detained, and after a few days he was again discharged.

December 16, 1885.—John Handforth has attended regularly as an out-patient since he was discharged from the Infirmary. During the seven or eight months he has been under observation his disorder of speech has undergone very little change. Asked to put out his tongue, he immediately opens his mouth, and on the demand being repeated, he points to his tongue with his index finger, but never protrudes it unless the examiner renders the request more specific by protruding his own. Asked to close his eyes he generally opens his mouth, but if the examiner should direct his eyes to the patient's, he alternately closes his eyes and opens his mouth in a confused and very comical manner.

To every question requiring a spoken answer, he replies in unintelligible jargon. For the last month or two there appears to be an occasional attempt at articulating a few simple words. Asked his name, for example, he utters two or three unintelligible syllables ending with "futh," which seems to be an attempt at pronouncing the last syllable of his name. Asked how he is getting on, an occasional "better" is heard amongst numerous unintelligible sounds. He is a very passionate man, and when he is vexed, which occurs but rarely in our presence, he gives vent to his feelings in a volley of very uncanonical language, which is very clearly articulated. And although the patient fails to comply with the most ordinary spoken request, such as "Close your eyes," and cannot give an intelligible answer to the simplest questions, he sometimes startles his examiner by giving manifest signs of having comprehended what appear to be rather complicated statements. One morning he was being examined along with William Davies, the case of motor aphasia already reported, and each of them seemed to be much interested in the case of the other. At the end of the examination I remarked to some students that the prospects of recovery were much worse in the case of John Handforth than in that of William Davies. I forget the exact words used, and I presume that I must have employed some expressive gestures, but, in any case, I had no expectation that the former would understand my remarks, although I knew the latter would. My surprise may, therefore, be easily imagined on finding that Handforth immediately raised both hands and shook his head, whilst, amongst some jargon, I thought I could detect, "Wus, Ah! Wus, Ah!" but whether that was so or not, his agitation and expressive gestures plainly indicated that he understood my remarks, and that what I had said had led him to believe that there was not much hope now left for his recovery. On another occasion I finished my examination of him, and, turning round, remarked to some students, in an off-hand, casual manner, "It is a very funny case." I immediately heard him puffing and snorting behind me, as he sometimes did when in a temper, and, on turning round, I at once saw that he was looking very indignant, while he kept repeating, "Fu fu fu-u-nny, damned fu-u-nny, humph!" When asked his age, he replies, in the usual jargon, in which the "f" sound frequently recurs, the attempt, probably, being to say fifty. When he is asked, "Are you a hundred years of age?" he does not deign to attempt a spoken reply, but he pouts his lips, closes his eyes, arches his eyebrows, and jerks his head slightly backwards and to the left, and, with a slight, expressive wave of the right hand, and shrug of the shoulders, he intimates, as plainly as he could do in spoken language, "If you believe me to be a fool, you never were more mistaken."

On being asked if he can read he gives vent to some gibberish, and by one or two very expressive gestures he indicates that he can do so quite easily. He, indeed, assumes a very airy and off-hand manner on such occasions, just as if he meant to convey the impression that reading is no trouble at all to him, and when the printed page is presented upside down to him he gives a disdainful snort, and immediately turns it right side up. He now places the printed paper upon a table, and points with the tip of the index finger of the left hand (the right being paralysed) to the paragraph to be read, and carrying his finger along each line in succession he gives vent to such a medley of unintelligible sounds that it is impossible for any one to note them down. During this pretended reading his face assumes a serious air, and the sounds are uttered with variations of tone and stops, although these do not appear to depend in any way on the sense or the punctuation of the printed page. When his finger points to the last word of the paragraph, he puts down the newspaper or book from which he was reading in the same airy, off-hand manner he assumed when asked to read, a manner which seems to say, "You see how very easy all this is to me."

When asked to read slowly, he immediately complies with the request, and pointing with his index finger to each word separately, he at the same time utters sounds which can be readily noted down by the observer. On November 25th, 1885, he was asked to read the following paragraph from a newspaper:—"The Bulgarian troops, after passing the night of the 22nd inst. on the positions taken from the Servians, commenced pursuing the enemy yesterday morning, and at four o'clock in the afternoon occupied Tsaribrod. Prince Alexander entered Tsaribrod yesterday." He deliberately pointed with the tip of his index finger to each word in succession, and uttered the following sounds, which were noted down at the time: "Thou for then to visen to ven are the do am they for then thou are they men thro they thro fordis and the thovis are thro and thro the asun am amli thro the then am thro the thro are musen or the they are on bind to vers." During this exercise he assumed an air of as much seriousness and importance as if he were giving those around him their first lesson in reading a foreign language, and when he had finished he gave a toss to his head and a slight wave of the hand, as much as to say, "You who are listening to me will be a long time before you can read with as much precision and elegance as that." His wife says that at home he frequently looks over the columns of a newspaper, and she believes that he understands the printed page, inasmuch as he often points out to her paragraphs which could only be selected by one who knows what would be likely to interest her. On being asked to write his name, he seizes the pen about an inch and a half from the point, between the finger and thumb of his left hand, and moves it upwards and downwards without resting his hand or forearm on the table. He then writes without hesitation, in a running hand—

Considering the awkward manner in which he holds his pen the result is surprisingly good. He is unable to copy from a printed page, but when I wrote down

he wrote under it the following—

He copies triangles, circles, and other geometrical figures readily, but cannot, of course, name them. On being examined one morning, he

D

wrote his name, and returned the pen as usual to me; but almost immediately afterwards a sudden thought seemed to strike him, he clutched the pen out of my hand, quickly drew the annexed figure, and putting down the pen he pointed to the drawing with an air of great satisfaction. At first I was at a loss to understand what he meant, and he tried to explain by the usual unintelligible sounds, but of course this kind of explanation was not a success. He now pointed to the figure and then to several miscellaneous articles lying on the table, but some time elapsed before his meaning dawned upon me. After a time it occurred to me that the oblique lines at the top of the vertical line were intended to represent the feathers of a quill pen and that the imperfect circle below was meant for an ink bottle. I now took hold of the ink bottle, and placing the pen in it I pointed to the figure, and he smilingly nodded assent. Some weeks later I placed a pen in the ink bottle, and asked him to make a drawing of them. On these occasions I had always to supplement my spoken requests by expressive gestures. He now seized the pen, and drew a figure, which might almost have been an exact copy of his former one, and he even took considerable pains to put the hook on the lower end of the vertical line, but what he meant by this hook I am still unable to understand, unless it be a device to prevent the pen from falling out of the bottle.

This case must bring to a close what we have to say with regard to the clinical varieties of aphasia, and we shall next proceed to make a few remarks on the morbid anatomy of this interesting affection.

4.—Morbid Anatomy.

In considering the morbid anatomy of aphasia we shall briefly discuss (1) the *nature* and (2) the *localisation* of the lesion.

(1.) *The Nature of the Lesion.*—Aphasia may be caused by (a) a functional or (b) an organic lesion of the brain.

(a) The most common *functional* lesion of the brain that gives rise to aphasia is the condition which precedes or follows an epileptic attack.[1] In some cases the warning of an epileptic attack consists of a sudden inability to speak,[2] and it is very probable that word-deafness and word-

---

[1] See Bernard (D.) "De l'aphasie et de ses diverses formes."—*Thèse de Paris*, 1885, p. 256; and Hughlings-Jackson (J.) "Clinical remarks on temporary loss of speech, and of power of expression (epileptic aphemia? aphrasia? aphasia?), and on epilepsics."—*Medical Times and Gazette*, Vol. I., 1886, p. 442.

[2] Moreau. "Aphasie (aphémie)."—*Gaz. des Hôp.*, Tome XXXVII., 1864, p. 70.

blindness is not by any means an uncommon aura, although the presence of these conditions is not so readily recognised by the observer as a motor aphasia, and in the subsequent descriptions of the patient they can hardly be distinguished from the mental confusion which attends the onset of unconsciousness. A transient aphasia often succeeds to the unilateral convulsions which are caused by gross lesions of the brain, such as gliomatous and gummatous tumours, and projecting spicula of bone. If the lesion is situated in one or other of the speech cortical centres themselves, then the aphasia is caused by an organic disease, and the symptom becomes permanent. In the cases under consideration at present, however, the lesion is situated at some little distance from any of the speech centres, but it gives rise to an irritation of one or other of them, which causes it to discharge its energy periodically, each such discharge being followed by exhaustion of the centre with loss of function until its energy is restored by nourishment. An embarrassment of speech, which in some cases may amount to a complete motor aphasia, not unfrequently ushers in an attack of migraine, and patients suffering from cerebral neurasthenia experience much difficulty in recalling names, especially proper names, while in aggravated cases there is a decided embarrassment of the faculty of expression which amounts to a minor degree of motor aphasia. A transient aphasia may also occur after a paroxysm of rage,[1] or after a severe fright,[2] shock,[3] or sunstroke,[4] and loss of the faculty of speech is also a well-recognised symptom of hysteria,[5] while in children and impressionable subjects an attack may be determined by the presence of intestinal worms or impacted fæces.[6] It is also met with as a complication of chorea.[7] Aphasia is also liable to supervene in the course of the eruptive and continued fevers, intermittent fevers,[8] diabetes,[9] gout, albuminuria and uræmia,[10] and saturnism,[11]

---

[1] Hovel (D. de B.) "On emotional aphasia."—*Medical Press and Circular*, Vol. II., 1875, p. 202.

[2] Habershon. "Case of aphasia consequent on fright."—*The Lancet*, Vol. II., 1870, p. 402.

[3] Midavaine. "Perte subite mais momentanée de la parole, suite d'un soufflet sur la joue."—*Arch. de la méd. Belge*, Tome VI., Brux., 1841, p. 243.

[4] Porter (P. B.) "A case of sunstroke followed by aphasia; death."—*Virginia Medical Monthly*, Richmond, Vol. I., 1874, p. 421.

[5] Davis (H. L.) "De l'aphasie hystérique." Paris, 1884.

[6] Gibson (D.) "Paralysis, with loss of speech, from intestinal irritation."—*The Lancet*, Vol. II., 1862, p. 139; and Mattei.—*Bull. de Acad. de Med.*, Tome XXX., 1864-5, p. 870.

[7] Jackson (J. B.) "Aphasia complicated with chorea."—*The Medical Press and Circular*, Vol. XXXVIII., 1884, p. 218.

[8] Boisseau (M. E.) "Aphasie transitoire liée à des accès de fièvre intermittente."—*Gaz. hebd. de Méd.*, 2, S., Tome VIII., 1871, p. 200.

[9] Bernard (D.) et Féré (Ch.) "Des troubles nerveux chez les diabétiques."—*Archiv de Neurologie*, Tome IV., 1882, p. 340.

[10] Hirtz.—*Gaz. méd. de Strasbourg*, 1865.

[11] Heymann (C.) "Ein Fall von Aphasia saturnina."—*Berl. klin. Wochenschrift*, Bd. II., 1865, pp. 195, 208, 223.

but as these diseases produce profound changes in the blood and blood-vessels, the disorder of speech is probably caused, not by a functional, but by one or other of the organic changes about to be described. The fact, however, that aphasia may be caused by the venom of serpents[1] shows that it may sometimes result from a poison circulating in the blood,[2] and aphasia has been observed in cases of chronic albuminuria in which no lesion could be discovered after death to account for the disorder of speech.

(b) The *organic* lesions which occasion aphasia are all those that produce destructive changes in the brain. The lesion may consist of a depressed portion of bone. The case of a gentleman is reported by Simon[3] who became suddenly unable to speak after a fall from his horse, without being paralysed. A small wound with depression was discovered over the left temple, and, death having occurred from meningitis, a piece of bone, which had been detached from the inner table, was found in the left third frontal convolution. More or less similar cases have been reported by Sydney Jones[4], MacCormack[5], Boyer[6], Decoudin,[7] Oulmont,[8] Mayor,[9] Déjérnie,[10] and others. Foreign bodies penetrating within the cranial cavity may be the cause of a permanent aphasia. I have myself met with a case in which a permanent motor aphasia, without paralysis, was caused by a gunshot wound in which the bullet entered the skull in the left temple and passed out at the top of the skull on the same side; and a large number of similar instances might be adduced. Intracranial growths of all kinds may, by compressing the central mechanisms of speech, give rise to any of the clinical varieties of aphasia, and every form of inflammation of the brain and of its membranes may also occasion this disorder of speech. Hæmorrhage of the brain may cause aphasia, but it is not by any means a frequent cause of the persistent forms of the affection. The reason of this is that the profounder forms of aphasia are caused by disease of the cortex of the brain, and a circumscribed hæmorrhage of the surface of

---

[1] Ogle (J. W.) "Loss of speech from the bite of venomous snakes."—*St. George's Hospital Reports*, Vol. III., 1868, p. 167.

[2] Dunoyer. "Aphasie transitoire toxique."—*Gaz. méd de Paris*, 1, S., 1884, Tome I., p. 461.

[3] Simon. "Casuistiche Beiträge zur Lehrevon der Aphasie."—*Ber. klin. Wochenschrift*, Bd. VIII., 1878, p. 959.

[4] Jones (Sydney).—*The Lancet*, Vol. II., 1873, p. 449.

[5] MacCormack.—*Brain*, Vol. I., 1877, p. 256.

[6] Boyer. "Études cliniques sur les lésions corticales des hémisphères cérébraux."—Paris, 1879.

[7] Decoudin.—*Bull. de la Soc. Anatomique*, Oct., 1876.

[8] Oulmont.—*Bull. de la Soc. Anat.*, Avril, 1877.

[9] Mayor.—*Bull. de la Soc. Anat.*, Julliet, 1876.

[10] Déjérnie.—*Bull. de la Soc. Anat.*, Janvier, 1879.

the brain is a comparatively rare accident. An example of aphasia from hæmorrhage is, however, reported by Rosenstein[1]. The case was that of a girl, aged twelve years, who had suffered from nephritis, and who was suddenly attacked with aphasia and agraphia a few weeks 'before the fatal issue, and after death a hæmorrhagic softening of the size of a hazel nut was found in the third left frontal convolution. Of all the lesions which give rise to aphasia occlusion of the Sylvian artery is by far the most frequent. The vessel may be occluded by an embolus, the plug being washed off from a diseased valve of the heart, or from an aorta whose internal surface is roughened by atheroma or injured by aneurism, or by thrombosis, the plug being formed by the deposit of fibrin upon the internal surface of a vessel deformed by atheroma, or by syphilitic endarteritis. It is probable that the aphasia which occurs in the course of the continued and eruptive fevers is caused, sometimes at least, by occlusion of some of the capillaries of the Sylvian artery by means of emigrant leucocytes; while the aphasia which occurs in the course of albuminuria and saturnism is likely to be caused sometimes by capillary, and at other times by massive hæmorrhages.

(2.) *The Localisation of the Lesion.*—It has just been seen that occlusion of one or other of the branches of the Sylvian artery is by far the most frequent cause of all forms of persistent aphasia, or of those from which recovery takes place but slowly. It is therefore desirable, b:fore proceeding to discuss the localisation of the lesion, to describe briefly the distribution of the branches of this artery.

The middle cerebral or Sylvian artery is the largest and most important branch of the internal carotid artery. As it passes outwards at the base of the brain it gives off small branches (Fig. 2, P)—the perforating branches, or the antero-lateral group of the ganglionic system of arteries—which pierce the anterior perforated space, and supply the corpus striatum and the anterior part of the optic thalamus. The following branches may be distinguished :[2]—(a) The *lenticular branches*, consisting of two or three small twigs, which ascend vertically, and enter the substance of the lenticular nucleus, and are distributed to its two inner divisions and to the adjoining portion of the caudate nucleus (Fig. 1, 5). (b) The *lenticulo-striate branch*, which is much larger than

[1] Rosenstein. "Diffuse nephritis. Urämische Anfälle von Temperaturerhöhung begleitet Aphasie als Terminalerscheinungen. Haselnussgrosses Blutcoagulum mit secondärer Erweichung in der dritten Stirnwindung der linken Hemisphäre."—*Berl. klin. Wochenschrift.*, Bd. V., 1868, p. 182.

[2] Duret (H.). "Recherches anatomiques sur la circulation de l'encéphale."—*Arch. de Physiologie.* 2me Série, Tome I. 1874. Pages 60, 316, 664, et 919. See also Heubner, "Zur Topographie der Ernährungsgebiete der Einzelnen Hirnarterien."—*Centralblatt f. d. med. Wissenschaften*, 1872, page 817.

any of the lenticular branches. It ascends along the external surface of the outer division of the lenticular nucleus, traverses the superior part of the internal capsule, and then passes from behind forwards into the substance of the caudate nucleus. It gives branches to the outer division of the lenticular nucleus, the internal capsules, and the caudate nucleus (Fig. 1, 4). (c) The *lenticulo-optic branch* passes, like the lenticulo-striate artery, along the external surface of the outer division of the lenticular nucleus, and through the posterior part of the internal capsule, and terminates in the anterior or external part of the optic thalamus.

On reaching the outer surface of the brain, the main trunk of the Sylvian artery divides into cortical branches. (a) The *first, or inferior frontal branch* (Fig. 2, 1) is limited in its distribution to the outer part of the orbital surface and the adjacent inferior or third frontal

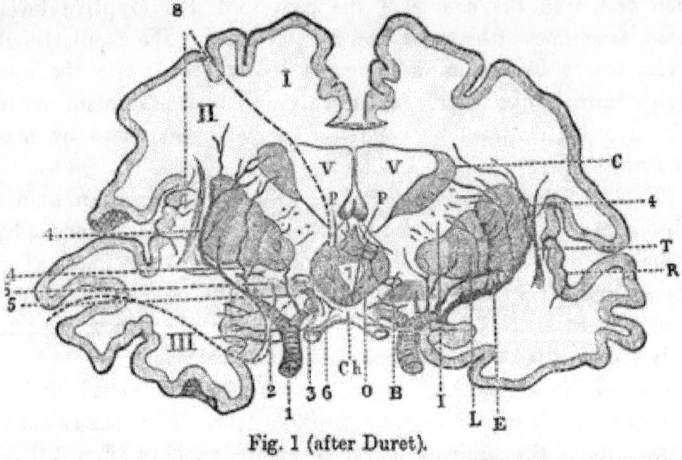

Fig. 1 (after Duret).

Transverse section of the cerebral hemispheres, about 1cm. behind the optic commissure arteries of the corpus striatum.

Ch. Chiasma; B—Section of the optic tract; L—Lenticular nucleus; I—Internal capsule; C—Caudate nucleus; E—External capsule; T—Claustrum; R—Island of Riel; V, V—Section of the lateral ventricle; P, P—Anterior pillars of the fornix; O—Grey substance of the third ventricle. *Vascular areas:* I.—Anterior cerebral artery; II.—Middle cerebral artery; III.—Posterior cerebral artery; 1—Internal carotid artery; 2—Sylvian artery; 3—Anterior cerebral artery; 4,4—Internal arteries of the corpus striatum (lenticulo-striato artery); 5, 5—Internal arteries of the corpus striatum (lenticular arteries). The lenticulo-optic artery is not represented in the figure.

convolution. (b) The *second, or ascending frontal branch* (Fig. 2, 2) supplies the posterior part of the middle frontal and the chief part of the ascending frontal convolutions. (c) The *third, or ascending parietal artery* (Fig. 2, 3) passes into the fissure of Rolando, and supplies the rest of the ascending frontal and the ascending parietal convolutions, as well as the anterior part of the superior parietal lobule. (d) The *fourth, or parieto-sphenoidal branches* (Fig. 2, 4) supply the inferior parietal lobule. The *fifth, or sphenoidal branches* (Fig. 2, 5) supply the superior tempero-sphenoidal convolutions.

The ganglionic branches of the Sylvian artery are all terminal arteries, and when one of these is occluded the part which it supplies inevitably undergoes softening. The cortical branches, however, are not terminal, but the anastomoses between the different branches of the artery itself, and between its branches and those of the anterior and posterior cerebral arteries are not free, and consequently when one of these branches becomes occluded some degree of softening generally takes place—we say generally, because in young people, in whom the arteries are very elastic, the main trunk of the Sylvian artery may be occluded immediately after the ganglionic arteries are given off without giving rise to softening. The case of a young girl came under my own observation, who, while

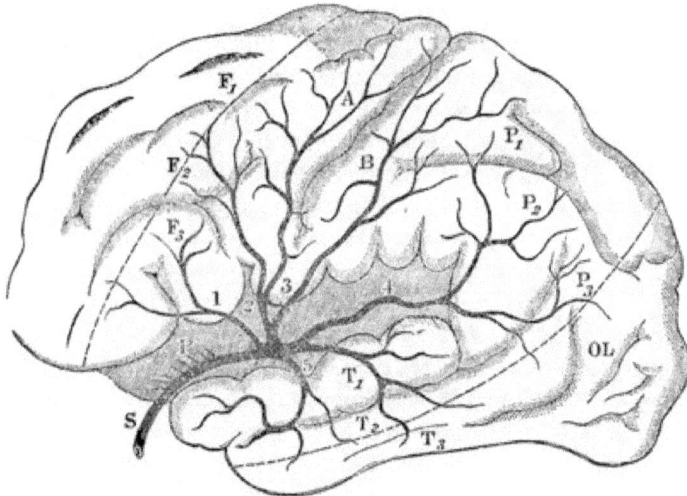

Fig. 2 (after Duret).

Diagram showing the area of distribution of the middle cerebral artery.

S—Sylvian or middle cerebral artery; P—Perforating branches; 1—Inferior frontal branch; 2—Ascending frontal branch; 3—Ascending parietal branch; 4—Parieto-sphenoidal branch; 5—Sphenoidal branches; A—Ascending frontal convolution; B—Ascending parietal convolution; F_1, F_2, F_3—First, second, and third frontal convolutions; P_1, P_2, P_3—First, second, and third parietal convolutions; T_1, T_2, T_3—First, second, and third temporo-sphenoidal convolutions. OL—Occipital lobe.

suffering from endocarditis, had first embolus of the femoral artery, and, in succession, embolus of the kidneys, spleen, and left Sylvian artery, occlusion of the latter being evinced by an attack of right-sided hemiplegia and combined motor and sensory aphasia. The patient having died from exhaustion about a fortnight after the attack of aphasia, an embolus was found in the main trunk of the Sylvian artery, a little outside the anterior perforated space, but before any of the cortical branches had been given off. Not the smallest speck of softening could be discovered, although it is very likely that the nerve cells in the area supplied by the

Sylvian artery had undergone fatty degeneration. There can be little doubt that, had the patient lived, the nutritive and functional activity of the tissues in this area would have been gradually restored. But in most cases of even young people, and in all cases in which the arteries have lost the elasticity of youth, occlusion of the Sylvian artery or of one of its branches is followed by a softening of greater or less extent in the area supplied by it; and when the embolus is lodged in the artery before the ganglionic arteries are given off, it is inevitable that the greater mass of the middle of the hemisphere must undergo rapid softening, and such cases are generally, if not always, rapidly fatal. Having advanced these anatomical considerations, we are now in a position to examine more narrowly into the localisation of the lesion in cases of aphasia, and to appreciate the relation which subsists between the position of the lesion in the different forms of this disorder and the distribution of the various branches of the Sylvian artery.

The faculty of language was placed by Gall in the supra-orbital lobes of the brain, but the first serious attempt to determine the localisation of this faculty by means of morbid anatomy was made by Bouillaud,[1] who, in 1825, came to the conclusion that speech was the result of the activity of the anterior cerebral lobes. In 1836 Marc Dax[2] advanced the opinion that the organ of language was situated in the left hemisphere of the brain, near to the Island of Reil, but this opinion had been neglected and almost forgotten until 1861, when Broca[3] localised the faculty of language in the third left frontal convolution. In the subsequent two years Broca[4] collected seventeen cases of aphasia in which post-mortem examinations had been obtained, and out of these the lesion was situated sixteen times in the posterior part of the third left frontal convolution, and once in the temporal lobe and Island of Reil. Soon afterwards cases were published by Parrot,[5] Fernet,[6] and Charcot,[7] in which a destructive lesion of the posterior part of the third right frontal convolution had not given rise to any disorder of speech, these cases affording the negative side of the proof of the assertion that the faculty of language is organised in the left hemisphere; and to make this proof still more striking, cases

[1] Bouillaud. "Traite d'encéphalité," 1825, p. 281.

[2] Dax (Marc). "Lésion de la moitié gauche de l'encéphale, coincidant avec l'oubli des signes de la pensée."—Montpellier, 1836.

[3] Broca (P.). "Sur le siège de la faculté de langage articulé, avec deux observations d'aphémie." Bull. de la Soc. Anat., Tome V., Août, 1861.

[4] Broca (P.). "Remarques sur le siège, le diagnostik, et la nature de l'aphémie."—Bull. de la Soc. Anat., Julliet, 1863.

[5] Parrot.—Bull. de la Soc. Anat., 1863, p. 372.

[6] Fernet.—Bull. de la Soc. Anat., 1863.

[7] Charcot (J. M.). "Sur une nouvelle observation d'aphémie."—Gaz. hebd. de Méd., Tome X., 1863, p. 473, et seq.

57

were reported by Voisin[1] and others, in which patients, who, on being the subjects of left-sided hemiplegia retained their speech, had become aphasic when subsequently attacked with a right-sided hemiplegia. In reproducing the work of his father in 1863, and again in 1878, M. Dax[2] fils, collected 371 cases of various diseases of the brain, chiefly from the works of Bouillaud and Lallemand. In 87 of these cases a lesion was found in the left hemisphere, and in all of them there was some disorder of speech during life; while in 53 cases a lesion was found in the right hemisphere, but disorder of speech was not found in any of them. The lesion was found in the left hemisphere 243 times out of 260 cases of hemiplegia with aphasia, collected by Seguin[3], and 140 times out of 146 cases collected by Voisin[4]. It was found by Callender and Kirkes that aphasia only failed once out of 13 cases of right hemiplegia, whilst it was only met with once in 13 cases of left hemiplegia. Lohnmeyer[5] collected 53 cases of aphasia in which an autopsy had been obtained, and out of these the lesion was situated 50 times in the left hemisphere; it was localised 24 times in the third left frontal convolution, and 24 times in it and the neighbouring parts. The Island of Reil was alone affected six times, the anterior portion of the frontal lobe twice, the middle lobe near the fissure of Sylvius thrice, the middle and posterior lobes twice, and the posterior lobe four times. Of 34 cases of loss of speech reported by Hughlings-Jackson[6] the paralysis was observed 31 times on the right, and three times on the left side; and out of 25 cases collected by Ogle[7] the lesion was situated in all of them in the left hemisphere. The analyses of collected cases made by various authors have now placed beyond any possibility of doubt that, in by far the majority of instances, the faculty of speech is organised in the left hemisphere of the brain. At the same time it has emerged that in a comparatively small number of cases the aphasia is associated with left-sided hemiplegia, and consequently it was inferred that disease of the right hemisphere of the brain does sometimes give rise to the disorder of speech. Cases of left hemiplegia with aphasia have been recorded by

[1] Voisin.—*Gaz. des Hôp.*, 25 Janvier, 1868.

[2] Dax (G.). "L'aphasie."—Montpellier, 1878, p. 112.

[3] Seguin.—*Quarterly Journal of Psychological Medicine*, Jan., 1868.

[4] Voisin. Art. "Aphasie." "Nouveau dictionnaire de Méd. et de Chir. pratique."—Tome III., Paris, 1866, p. 4.

[5] Lohnmeyer. "Kann Aphasie zur Trepanation veranlassen?"—*Archiv f. klin. Chirurgie*, Bd. XIII., 1871-2. p. 309.

[6] Hughlings-Jackson (J.). "Loss of speech: its association with valvular disease of the heart, and with hemiplegia on the right side; defects of speech in chorea."—Clinical Lecture and Report, *Lond. Hosp. Reports*, Vol. I., 1864, p. 388.

[7] Ogle (J. W.). "Aphasia and agraphia."—*St. George's Hospital Reports*, Vol. II., 1867, p. 117.

Pye-Smith, Hughlings-Jackson[1], John Ogle[2], and Ferrier[3], and post-mortem evidence of aphasia being caused by disease of the right hemisphere of the brain was obtained by Wadham[4] and Habershon[5]. A negative proof of the occasional organisation of speech in the right hemisphere was afforded by F. Taylor[6] and Foulis[7], who reported cases of destruction of Broca's convolution in the left hemisphere without giving rise to aphasia. Care should be taken not to conclude too hastily that in all cases of left-sided hemiplegia with aphasia the lesion causing the disorder of speech is situated in the right hemisphere of the brain; inasmuch as cases have been recorded by Raymond and Dreyfous[8] in each of which a lesion was found in the right hemisphere, causing a left hemiplegia and another in the third left frontal convolution, which doubtless caused the aphasia. On the whole the observations collected and analysed by various authors prove beyond any possibility of doubt that aphasia is caused in by far the majority of cases by lesion of the left hemisphere, and in a comparatively few cases by lesion of the right hemisphere of the brain. We must now endeavour to localise more accurately the different forms of the affection.

In *motor aphasia* the lesion is most usually situated in the posterior part of the third left frontal convolution. Broca thought that all aphasic disorders of speech were caused by disease of this part of the brain, and the cases he adduces in favour of this opinion were, with one exception, in which the lesion was found in the temporal lobe and Island of Reil, all examples of pure motor aphasia. In the first case in which he obtained an autopsy the patient—Leborgne—understood everything that was said to him, but was only able to reply *tan, tan*, to all questions, and when his interlocutors failed to understand the lively gestures by

[1] Hughlings-Jackson (J.). "Loss of speech with hemiplegia of the right side; valvular disease; epileptiform seizure affecting the side paralysed."—*Med. Times and Gazette*, Vol. II., 1864, p. 166. "Aphasia with hemiplegia of the left side.—*The Lancet*, 1868, p. 316; and *Ibid*, Vol. I., 1868, p. 370.

[2] Ogle (John W.). "Illustration of the impairment of the power of intellectual language."—*The Lancet*, Vol. I., 1868, p. 370.

[3] Ferrier. "Left-sided epileptic hemiplegia and aphasia."—*The Lancet*, Vol. II., 1880, p. 730.

[4] Wadham (W.) "Case of hemiplegia of the left side in an ambidextrous boy; subsequent occurrence of aphasia which continued complete for three months; gradual but imperfect recovery of speech; nearly total destruction of the Island of Riel on the right side."—*St. George's Hospital Reports*, Lond., Vol. IV., 1869, p. 245.

[5] Habershon. "Aphasia, with hemiplegia on the left side, and tumour on the right side of the brain in the third frontal convolution."—*The Lancet*, Vol. II., 1880, p. 979.

[6] Taylor (F.) "Right hemiplegia after scarlatina, with embolism of the left middle cerebral artery; destruction of Bovia's convolution without aphasia."—*The Lancet*, Vol. II., 1880, p. 896.

[7] Foulis (D.) "A case in which there was destruction of the third left frontal convolution without aphasia."—*The British Medical Journal*, Vol. I., 1879, p. 383.

[8] Raymond et Dreyfous (F.) "Contribution a l'étude de l'aphasia."—*Archiv. de Neurologie*, Tome III., 1882, p. 80.

which these words were accompanied he gave vent to much unintelligible jargon, amidst which was clearly articulated, "Sacré nom de Dieu." The lesions found in the brain after death were widely distributed over the surface and in the substance of the left hemisphere, but Broca believed that destruction of the third frontal convolution was the cause of the disorder of speech. A few months later he had an opportunity of examining the brain of Lelong, a man who had the use of only five words, and in his case the lesion consisted of a cavity about the size of a franc piece filled with serous fluid, and which was almost strictly circumscribed to the posterior part of the third left frontal convolution, encroaching to a slight extent upon the posterior part of the second frontal. From these cases Broca, with his usual caution, only concluded that "the integrity of the third and probably of the second frontal convolution appears to be indispensable to the exercise of the faculty of articulate speech." In the subsequent two years Broca collected, as already mentioned, 17 cases of aphasia with post-mortem examinations, in 16 of which the lesion was situated in the third left frontal convolution, and once in the temporal lobe and Island of Reil. The cases of motor aphasia recorded in which the lesion was situated in the third left frontal convolution are now so numerous and so well known that it is quite unnecessary to quote any of them here. But although it is proved that lesion of the posterior part of the third left and in occasional cases of the third right frontal convolution gives rise to a motor aphasia, it does not follow that this disorder of speech may not be caused by disease of other parts of the brain. And, indeed, it has been found that lesions of the centrum ovale underlying the frontal convolution may give rise to a motor aphasia which is as complete and persistent as that caused by disease of the cortex itself. A man, aged 60 years, whose case is reported by Pitrés[1], suffered from right hemiplegia, with embarrassment and finally complete loss of speech. At the autopsy, two small patches of yellow softening were found in the cortex of the left hemisphere, one being situated in the superior parietal lobule, and the other in the angular gyrus. The third left frontal convolution was normal, but a large focus of softening was observed in the centrum ovale, which extended anteriorly to the part underlying the posterior extremity of the third frontal convolution, and posteriorly beyond the posterior extremity of the optic thalamus. It is possible that the spot of softening in the angular gyrus in this case might have caused word-blindness, but it could not have given rise to complete loss of speech, which, therefore, must have resulted from the softening in the

[1] Pitrés (A.) "Lésions du centre ovale." ("Obs." xxxviii.) Paris, 1877, p. 94, et seq.

centrum ovale. A case of aphasia is reported by Bouchard,[1] in which the lesion, a yellow softening, was situated in the anterior and superior part of the intraventricular portion of the left corpus striatum, which corresponds to the anterior segment of the internal capsule. A secondary degeneration was observed in the crusta and the anterior pyramid of the medulla oblongata of the same side. A case of aphasia is likewise reported by Farge,[2] in which the lesion was situated in the centrum ovale, immediately above and to the outer side of the caudate nucleus of the corpus striatum,[3] almost in the same position as the one reported by Pitrés. A case of right-sided hemiplegia, with aphasia, is reported by Hughlings-Jackson,[4] in which the lesion was found in the corpus striatum, and we shall hereafter find that most cases of ordinary right-sided hemiplegia, which are most probably caused by hæmorrhage into the lenticular nucleus of the corpus striatum, are accompanied by a transient motor aphasia, recovery taking place in a few weeks. A case in which a circumscribed hæmorrhage in the pons Varolii produced a slight degree of motor aphasia is reported by Hermann Weber and Altdoerfer.[5] A man, aged 35 years, was seized with sudden loss of consciousness and left hemiplegia. He had impairment of the power of swallowing and of articulation. He had difficulty in finding the right words, which, however, when suggested, were readily pronounced. The patient died three weeks after the attack, and at the autopsy a small hæmorrhagic spot was found in the anterior part of the right half of the pons Varolii, with softening of the surrounding substance.

Motor agraphia is generally associated with aphemia, and consequently it may be inferred that the motor cortical centre for the special movements of writing lies near to that for the special movements of spoken speech. The fact, however, that the disability of spoken speech and of writing are not always present in equal degree in cases of motor aphasia, and that in rare cases the one faculty may be lost while the other is retained, shows that these centres are not identical. The motor

[1] Bouchard (C.) "Aphasie sans lésion de la troisième circonvolution frontale gauche."—*Comptes Rendus des Séances de la Soc. de Biologie*, 1864. Paris, 1865. 4, S., Tome I., p. 111.

[2] Farge (E.). "Hémiplégie droite, et aphasie sans lésion de la troisième circonvolution frontal gauche."—*Gaz. hebd. de Méd.*, Paris, 1864. 2, S. Tome I., p. 724.

[3] See also Déjernie (J.). "Aphasie et hémiplégie droite ; disparition de l'aphasie au bout de neuf mois ; persistance de l'hémiplégie : mort par phthisie au bout de trois ans (intégrité de la troisième frontale, lésion du faisceau pédiculo-frontale inférieur, du noyau lenticulaire, et de la partie anterieure de la capsule interne."—*Progrès Méd.*, Paris, 1879, Tome VII., p. 468.

[4] Hughlings-Jackson (J.). "Disease of the left side of the brain, involving the corpus striatum, &c. Aphasia."—*Medical Times and Gazette*, Vol. II., 1884, p. 605.

[5] Weber (H.) and Altdoerfer. "Case of aphasia with left hemiplegia. Hæmorrhage and softening in the right side of the pons Varolii."—*The British Medical Journal*, Vol. I., 1877, p. 13. See also Wood, "Tumour at the base of the brain, pressing upon medulla spinalis."—*Phild. Med. Times*, 1881-2, Vol. XII., p. 648.

centre for writing is placed by Exner[1] in the posterior part of the second left frontal convolution. Of the four cases which he adduces in favour of this opinion, the most convincing is probably a case of agraphia reported by Bar,[2] in which the lesion, consisting of a clot of blood, occupied the posterior part of the second left frontal convolution. In a case of aphasia and agraphia reported by Nothnagel,[3] a spot of softening from embolus was found in the posterior part of the second, another in the corresponding part of the first, and three in the convolutions of the Island of Reil.

In *apperceptive aphasia* the lesion is localised in the area of distribution of the parieto-sphenoidal and sphenoidal branches (Fig. 2, 4 and 5) of the Sylvian artery, and the region of softening comprises the supra-marginal convolution, the angular gyrus (visual centre), the posterior part of the infra-marginal convolution (acoustic centre), and the convolution bounding the parallel and collateral fissures. In the cases of aphasia with autopsies collected by Broca and the earlier observers, the lesion was occasionally observed to be situated in the temporal lobe and the posterior part of the parietal lobe ; but these cases were regarded as exceptions to the general rule of localisation in the third frontal convolution, up to the time when Wernicke[4] reported a case of that disorder of speech, which Kussmaul[5] subsequently named word-deafness, and in which the lesion consisted of softening from thrombosis of the first and a large part of the second temporo-sphenoidal convolutions (Fig. 3). In the case of word-deafness reported by Giraudeau, already quoted at length, the lesion consisted of a tumour about the size of a nut, which was situated in the posterior part of the two first temporo-sphenoidal convolutions. A case of word-deafness is reported by Kahler and Pick,[6] and another by Fritsch,[7] in which the lesion was found in the first and second left temporo-sphenoidal convolutions of the left hemisphere.

In two cases of word-deafness, reported by Petrina, and in one

[1] Exner (Sigmund). "Untersuchungen über die Localisation der Functionen der Grosshirnrinde des Menschen."—Wien, 1881, p. 57.

[2] Bar.—*France Méd.*, 1878, No. 77.

[3] Nothnagel. "Topische Diagnostik der Gehirnkrank."—Berlin, 1879, p. 482.

[4] Wernicke. "Der aphasische Symptomencomplex."—Breslau, 1874, p. 19. See also *Lehrbuch der Gehirnkrankheiten für Aerzte und Studirende*, Bd. II., 1881, p. 556, et seq.

[5] Kussmaul. Art. "Disturbances of speech."—"Ziemssen's Cyclopedia," Vol. XIV., 1878, p. 770.

[6] Kahler and Pick. "Beitrag zur Lehre von der Localisation du Hirnfunctionen."—*Prager Vierteljahrschrift*, Bd. CXLI., 1879.

[7] Fritsch.—*Wiener Medizinische Presse*, 1880. See also Skwortzoff (Mlle. N.). "De la cécité et de la surdité des mots dans l'aphasie." Paris, 1881, p. 87.

[8] Petrina (Theodor). "Sensibilitätsstörungen bei Hirnrindläsionen."—*Zeitschr. f. Heilkunde*, Bd. II., 1881. p. 373. Abstr. *Neurolog. Centralbl.*, Bd. I., 1882, p. 12.

reported by Claus,[1] the lesion was found in the anterior part of the superior temporo-sphenoidal convolution. A very remarkable case of apperceptive aphasia was communicated by Wernicke[2] to the Berlin Physiological Society in 1883. The patient was first affected by word-deafness, and at a subsequent period he became completely deaf; he died from an intercurrent disease, and a spot of softening was found in the first temporo-sphenoidal convolution in each hemisphere of the brain. A careful dissection proved the absence of any local disease in the peripheral organ of hearing. A case is reported by Pitrés[3] in which the patient, who was the subject of an old left-sided hemiplegia, was seized by a second apoplectic attack, but without paralysis, and although he appeared to take considerable notice of what passed around him, he only replied by an unintelligible grunt on being loudly asked his name.

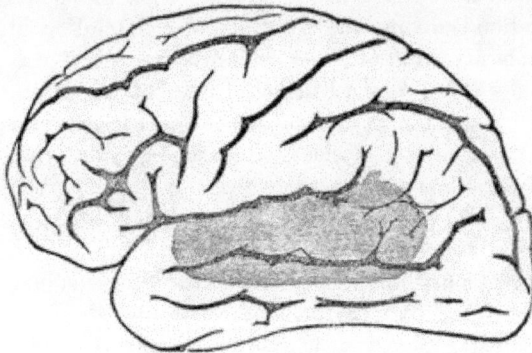

Fig. 3.

He died on the second day after the last attack, and at the autopsy, besides old lesions in the right hemisphere, a recent clot of blood, about the size of a nut, was found in the medullary substance of the temporo-sphenoidal convolutions.

A somewhat complicated disorder of speech is reported by Broadbent[4], but a very prominent feature of the case was that the patient did not understand spoken or written speech, while his attempts at expressing his own thoughts ended in mere jargon. The patient died suddenly, and on post-mortem examination, softening was found, involving a considerable part

---

[1] Claus. "Zur Casuistik der Localisation der Grossbirnfunctionen."—*Irrenfreund*, 1883, No. 6, Abstr. *Neurolog. Centralbl.*, Bd. III., 1884, p. 205.

[2] Wernicke.—*Sitzung der Berliner physiologische Gesellschaft*, Marz 9, 1883. Abstr. "*Nature*," Vol. XXVII., 1883, p. 503.

[3] Pitrés (A.). "Recherches sur les lésions du centre ovale des hémisphères cérébraux."— Paris, 1877 (Obs. XV.), p. 55.

[4] Broadbent (W. H.) "On a case of aneurism, with post-mortem examination."—*Medico-Chirurgical Transactions*, Vol. LXI., 1878, p. 147; and *The Lancet*, Vol. I., 1878, p. 312.

of the convex surface of the left hemisphere, the convolutions affected being the supra-marginal lobule, the angular gyrus, the first and second temporo-sphenoidal convolutions, and the adjacent part of the occipital lobe (Fig. 4). It is certain that in this case the patient suffered both from word-blindness and word-deafness. A case of combined word-deafness and word-blindness is reported by Rosenthal,[1] which was caused by destruction of the posterior third of the first and second temporo-sphenoidal convolutions, the lesion extending to the point of junction of these convolutions, and the angular gyrus and inferior parietal lobule. A case of partial word-blindness is reported by Déjerine,[2] in which the patient could read, but did not understand what she read. The autopsy revealed the existence of a tumour, about the size of an orange, in the region of the inferior parietal lobule. The case of a woman is reported by Dr. Shaw,[3] of Brooklyn, who had suddenly become, after a slight apoplectic attack, aphasic and perfectly deaf and blind. At the autopsy

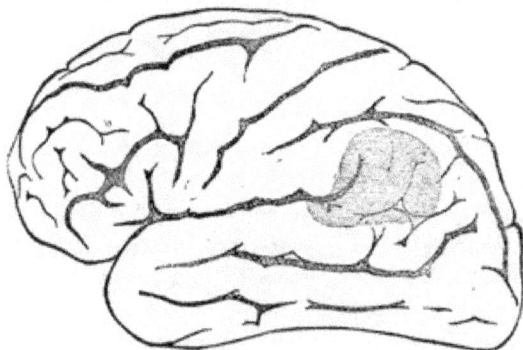

Fig. 4.

complete atrophy of the angular gyri of both hemispheres was found, and no other lesion either in the brain or peripheral organs. Most cases of word-blindness are associated with right-sided bilateral hemianopsia, and it is now known that this form of hemianopsia[4] may be caused by lesion of the cortex of the angular gyrus and neighbouring parts, or of the

---

[1] Rosenthal (Warschau). "Ein Fall von corticaler Hemiplegie mit Worttaubheit."—*Central-blatt für Nervenheilkunde*, 1884, No. 1.

[2] Déjerine (G.). *Bull. Soc. de Biologic*, 7. S. Tome II., p. 261. See *Progrès Méd.*, Tome VIII., 1881, p. 629.

[3] Shaw (J. C.). *Archives of Medicine*, Feb., 1882. Abstr. *Brain*, Vol. V., 1882-3, p. 430.

[4] See Hirschberg. *Deutsche Zeitschrift für prakt. Med.*, 1876, Nos. 4 u. 5. Westphal, "Zur Localisation der Hemianopsie und des Muskelgefühls beim menschen."—*Charité-Annalen*. Jahrg. VII. (1880), Berl., 1882, p. 466. Marchand, "Beitrag zur Kenntniss der homonymen bilateralen Hemianopsie und der Faserkreuzung im Chiasma opticum."—*Archiv. f. Ophth.*, Bd. XXVIII., Abstr. 2, 1882, p. 63 ; and Willbrand, "Ueber Hemianopsie und ihr Verhältniss zur topischen Diagnose der Gehirnkrankheiten." Berlin, 1881.

subjacent white substance of the left hemisphere. In most of the cases in which an autopsy has been obtained, the hemianopsia is said to have been accompanied by aphasia, and although the clinical description of the speech disorder is not sufficiently minute to enable us to judge of the form of aphasia present, it may be inferred that it was word-blindness.

In combined motor and sensory aphasia the lesion is widely distributed over the surface of the brain, or two separate lesions may be present. A case is reported by Boyer,[1] in which the patient had suffered for three years from right-sided hemiplegia with contractures, aphasia, and deafness. At the autopsy a spot of softening was found in the posterior part of the first and second frontal convolutions, and another in the convolutions bordering the horizontal limb of the Sylvian fissure. The acoustic and hypoglossal nerves and their nuclei in the medulla were found healthy.

A very remarkable case of right-sided hemiplegia with contracture, associated with aphemia, word-blindness, and partial word-deafness is recorded by Bernard.[2] At the autopsy numerous spots of softening were found in the left third frontal, the inferior part of the ascending frontal, and the middle portion of the ascending parietal convolutions. The softening had extended into the substance of the hemisphere and destroyed the anterior segment of the internal capsule. The middle part of the ascending parietal convolution, the angular gyrus, and the posterior part of the first tempero-sphenoidal convolution were occupied by a patch of yellow softening. Another patch of softening was found at the bottom of the sulcus of Rolando below its middle, and the surface of the posterior part of the Island of Reil was destroyed. It is probable that in many cases of verbal amnesia the lesion consists of great local anæmia of the cortex from degenerated arteries, without any destructive disease. A case of verbal amnesia without word-deafness, however, is reported by Rosenthal,[3] in a patient suffering from general paralysis ; this defect of speech was ushered in by an apoplectiform attack, and persisted unchanged for upwards of two years. At the autopsy, besides evidence of a chronic leptomeningitis, an old focus of softening was found in the second and third temporo-sphenoidal convolutions, the first temporo-sphenoidal convolution being quite free from disease.

If we now summarise the facts collected in the foregoing pages it will be seen that, in by far the majority of cases, aphasia is caused by a lesion

---

[1] Boyer (Cl. de). "Etude sur les lésions corticales des hémisphères cérébrales." Paris, 1879, p. 91.

[2] Bernard (D.). "De l'aphasie et de ses diverses formes."—Paris, 1885, p. 244.

[3] Rosenthal (Albert). "Allgemeine Paralyse, mit sensorischer Aphasie amrürt."—*Centralblatt für Nervenheilkunde*, Jahrg. IX., April 15, 1886, p. 225.

of the left hemisphere of the brain, and that the lesion is situated in aphemia in the posterior part of the third frontal convolution (Fig. 5, F_3); in agraphia, probably in the posterior part of the second frontal and the adjoining part of the ascending frontal convolutions (Fig. 5, F^2); in word-deafness, in the first and second temporo-sphenoidal convolutions (Fig. 5, T^1); and in word-blindness, in the angular gyrus and adjoining part of the inferior parietal lobule. It would appear from the case just quoted from Rosenthal that the aphasia of recollection, or verbal amnesia, is caused by lesion of the second and third temporo-sphenoidal convolutions (Fig. 5, T^2 and T^3). It has already been stated that occlusion of one or other of the branches of the Sylvian artery is the most frequent cause of aphasia, and if we express the morbid

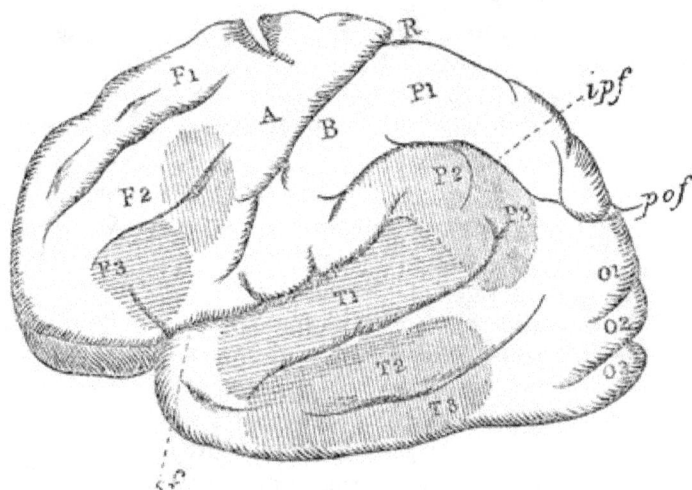

Fig. 5.

External surface of the left hemisphere of the Brain.—Fissures.

R—Fissure of Rolando; Sf—Fissure of Sylvius; ipf—Interparietal fissure; pof—External parieto-occipital fissure; F_1, F_2, and F_3—The first, second, and third frontal convolutions; A and B—The ascending frontal and parietal convolutions respectively; T_1, T_2, and T_3—The first, second, and third temporo-sphenoidal convolutions; O_1, O_2, and O_3—The first, second, and third occipital convolutions; P_1—The superior parietal lobule; P_2—The inferior parietal lobule; P_3—The angular gyrus. The shaded part on F_3 corresponds to the localisation of the lesion in *aphemia*; on F_2 to *motor ographia*; on T_1 to *word-deafness*; on T_2 and T_3 to *verbal amnesia*; and on P_2 and P_3 to *word-blindness*.

anatomy of this affection in terms of the arterial supply of the cortex, it may be said that occlusion of the inferior frontal branch gives rise to aphemia, of the anterior division of the ascending frontal branch to agraphia, of the sphenoidal branches, generally the posterior division, to

E

word-deafness, and of the parieto-sphenoidal branch to word-blindness (Fig. 2). It must not, however, be imagined that the branches of the Sylvian artery are invariably separated so clearly from one another in nature as they are in our diagram.

The fact, for instance, that aphemia and motor agraphia are so frequently associated, seems to show that the centres for the regulation of articulate speech and for writing must not only lie near together, but be also supplied by the same vascular system, and it is therefore probable that the inferior frontal, and the anterior division of the ascending frontal branch are not so trenchantly separated from one another as they are in the diagram. It must also be remembered that the spot of softening which occurs after the occlusion of the main trunk of the Sylvian artery, or of one of its branches, is by no means co-extensive, especially in young people, with the area of distribution of the vessel. William Davies (Case 1), for instance, suffered from a right-sided hemiplegia and complete sensory and motor aphasia, and taking the symptoms in association with the fact that he had a systolic murmur over the apex of the heart, it is very likely that the main trunk of the Sylvian artery was occluded by an embolus immediately after the perforating branches are given off. In this case, however, the patient made a comparatively rapid recovery from the sensory aphasia and from the hemiplegia, and even the faculty of writing has undergone considerable improvement, but he is still, twelve months from the onset of the attack, the subject of almost complete aphemia; and it may consequently be inferred that the posterior part of the third, and to a less extent the posterior part of the second, and the adjoining part of the ascending frontal convolution was the only part of the cortex which had become permanently disorganised. In combined sensory and motor aphasia which becomes persistent, the softening is widely distributed in the area of distribution of the anterior and posterior branches of the Sylvian artery. Such cases usually occur in persons advanced in life, and the aphasia is generally associated with hemiplegia and permanent contractures. It is possible for a pure motor or a pure sensory aphasia to be associated with persistent hemiplegia, but it is probable that in such cases the aphasia is caused by a circumscribed lesion in the cortex, and the hemiplegia by a second lesion, which injures the pyramidal tract in its passage through the internal capsule, crus cerebri, or pons. When, however, a persistent hemiplegia and aphasia are caused by one and the same lesion, the disorder of speech is likely to implicate both the apperceptive and the motor functions, and the disease probably extends widely over the greater part of the area supplied by the branches of the Sylvian artery.

5. Morbid Physiology.

It now remains for us to interpret the symptoms of aphasia by means of the knowledge we possess of the physiology of those portions of the cerebrum which have been found diseased in the different clinical varieties of the affection. We have already been led to the conclusion that loss of the expressive faculty of speech is, speaking broadly, caused by disease of the motor area of the cortex of the left hemisphere, and that loss of the impressive or apperceptive faculty is caused by disease of the sensory area of the same hemisphere. But in order to justify this conclusion, and to make our interpretation more detailed and minute, it is desirable to subject the symptoms of the different forms of aphasia to a still further analysis. And let us begin our examination with the aphasic disorders of the expressive faculty of speech.

Motor aphasia may, as we have already seen, be divided into (1) aphemia, (2) agraphia, and (3) amimia. It must at once be admitted that nothing is known of amimia as a separate affection, and not even sufficient is known of it in its association with other speech disorders to enable us to enter profitably upon a further analysis of it. Passing over amimia, therefore, aphemia and agraphia remain for consideration, and although these disorders are usually found combined in various degrees in the same person, it will be found useful to subject each to separate analysis.

(1.) *Aphemia* presents many degrees, from a slight defect in the power of expressive speech up to absolute wordlessness. In the *first* or the least degree of the disorder the patient has either partially recovered or has never altogether lost the expressive faculty of speech. The subject of aphemia who has been master of two or more languages may, for example, recover the use of one of these, while he remains totally unable to express himself in either of the others. The case of a Russian gentleman is reported by Charcot, who, besides his mother tongue, spoke French fluently, and German less readily. A friend addressed him in French, and he readily comprehended what was said to him, but found himself quite unable to reply in the same language. The German language he found difficult to understand, and was totally unable to speak in it. Numerous examples of this kind might be quoted from other authors. In some cases the disorder of speech is manifested by an uncertain and hesitating speech. Slowness of utterance may, indeed, of itself be taken as a sign of a slight degree of aphemia, just as slowness in the execution of the "devil's tattoo" on the table may be taken as a sign of partial paralysis of the fingers. In other cases the disability is declared by the misplacement or mutilation of words. The patient may, to quote the examples adduced by Dr. Hughlings Jackson, use a word kindred in its meaning with the one intended, as "worm powder" for "cough medicine," or in its sound, as "parasol" for "castor oil." It is, however, in sensory aphasia that the most marked forms of the misplacement of words are observed, and the condition will be subsequently described under the name of *paraphasia*.

In the *second* degree of aphemia the patients have lost the power of expressing their thoughts in spoken words with the exception of a few monosyllabic replies to simple questions. Such patients, however, are able to repeat most words uttered in their hearing, and they may be able to read aloud without committing any serious mistakes. A boy, who was the subject of aphemia, and to whose case we have already alluded, was in the Convalescent Hospital at Cheadle last summer. He could only utter spontaneously "Yes" and "No," his own name and a few other words imperfectly, but he could repeat a large number of monosyllabic names, such as horse, cow, keys, book, etc. When, however, the corresponding objects were shown to him he was unable to name them, but instantly recognised the right name, and repeated it, when it was uttered in his hearing. In other cases the power of repeating words and of reading aloud is retained or recovered in even higher degree than it was in the case of this boy, while the power of spontaneous vocal speech remains very limited.

In the *third* degree of aphemia, the power of repeating words or reading aloud is lost, but the patient can give monosyllabic replies to two or three simple questions. The unit of emotional language is, so far as it is expressed in vocalisation, an exclamation or an interjectional word or phrase, but the unit of intellectual language is a proposition, and if the monosyllabic reply of an aphemic patient possess any speech value it must be equivalent to a distinct proposition. If, for example, a patient says "Yes!" as an exclamation, the word belongs to emotional language ; but if he says "Yes," with the full understanding that he is giving an affirmative answer to a question, then the word is equivalent to a distinct proposition. Suppose the patient is asked "Are you better to-day?" and he replies "Yes," with a full comprehension of the meaning of his answer, it is equivalent to the proposition "I am better to-day," and the reply itself becomes part of intellectual language. When William Davies (Case 1) was asked his name some months after his attack, he answered "Bill," which is the equivalent of the proposition "My name is Bill," and therefore a part of intellectual language. In this degree of aphemia, therefore, the patients retain or recover the use of a few simple words which possess real speech value.

In the *fourth* degree of aphemia, the patient has completely lost the power of giving expression in vocal language to any thought, but although he is speechless he is not wordless. The words which the patient can use continue, as a rule, the same in the same patient, and consequently they have been named "recurring utterances." These utterances usually consist of such simple words as "Yes" or "No," but whatever the word may be it is used on all occasions, whether appropriate or not, and it consequently possesses no speech value. At other times the recurring utterances consist of a phrase, such as "Come on to me," or, "I want protection."[1] The man whose recurring utterance was "Come on to me," was a railway signalman, who had been taken ill on the rails in front of his box ; while the man who could only say "I want protection," had his left cerebral hemisphere injured in a brawl. Dr. . Hughlings-Jackson makes the very probable supposition that in these cases the recurring utterance constituted the last words spoken, or which were in a state of mental preparation for utterance, when the damage occurred to the brain. It is not improbable but that words uttered, or about to be uttered, during a period of great excitement might leave permanent traces on the organisation of the brain which would render them liable to be afterwards uttered as interjectional phrases during emotional states. Besides these recurring utterances, the patient may,[2] under certain circumstances, give vent to "occasional utterances." When the patient becomes vexed with his interlocutor for not understanding his recurring utterance, or at himself for his inability to express himself, he may find relief in a volley of oaths. On rare occasions a patient may even utter a phrase which is appropriate to the occasion, such as "Good-bye,"[3] when a friend is leaving. Now, of these occasional utterances, swearing is a purely emotional expression, and even the phrase "Good-bye" may be regarded as expressing a state of mental regret rather than an intellectual appreciation of the surrounding circumstances. That these recurring and occasional utterances must be regarded as belonging to emotional rather than to intellectual speech is shown by the fact that the patient is unable to repeat, when asked, his favourite oath, his formula of leave-taking, or the simple recurring utterance of "Yes" or "No." The few words which the patient uses may be accompanied by such variations of tone and gesture as indicate when he is angry or joyful. The words are indeed, in Mr. Herbert Spencer's[4] language, more akin to song than to speech, and belong to emotional and not to intellectual language.

[1] See Jackson (Hughlings). "On affections of speech from disease of the brain." *Brain*, Vol. I., 1878-9, p. 313 ; and Vol. II., 1879-80, p. 201.

[2] Jackson (Hughlings). "Remarks on the occasional utterances of 'speechless' patients."— *The Lancet*, Vol. II., 1867, p. 70.

[3] Broadbent (W. H.). "A case of peculiar affection of speech, with commentary." *Brain*, Vol. I., 1878-9, p. 494.

Spencer (Herbert). "The origin and function of music." Essays scientific, political, and speculative. Vol. I., 1868, p. 221.

In the *fifth* or highest degree of aphemia the patients, so far as vocal speech is concerned, are both speechless and wordless. Such patients may, however, endeavour to reply to questions by grunting noises in the throat, or by the continuous repetition of a meaningless syllable, like the "tan, tan" of the patient Leborgne, whose case was reported by Broca. The affection of speech, which has been named in these pages *jargon aphasia*, must, in so far as the expressive faculty is implicated, be included in this category. In these cases the patient, instead of continuously repeating the same syllable, gives utterance to a rapid succession of different syllables; but it may be noted that two, or at most three, favourite sounds frequently recur. This form of aphasia is, as already stated, frequently a compound disorder, and the expressive disability is then accompanied by a considerable disorder of the apperceptive faculty of speech.

(2) *Motor agraphia* may, like aphemia, be divided into five varieties, according to the degree of the affection.

In the *first* or least degree the patient either has acquired or has never completely lost the power of writing spontaneously a few words, phrases, and simple sentences, although he makes egregious mistakes in spelling and diction.

In the *second* degree he has lost the power of giving spontaneous expression to his thoughts, but he can write to dictation, and copy from a written or printed page. The power of writing to dictation varies greatly, and is often very limited and imperfect. The capacity to copy also varies within extreme limits; the patient at first may only be able to copy slowly and imperfectly a written word in written and a printed word in printed characters, but after a time he may reacquire the power of writing in a running hand from a printed page.

In the *third* degree of agraphia the patient is unable to write to dictation or to copy sentences, but he can still write somewhat imperfectly his own name, and probably a few simple words, especially if aided by a plain copy of them.

In the *fourth* degree, or what Kussmaul calls *verbal* agraphia, the patient can write rows of letters, which may be separated here and there in the form of words, but which do not convey any meaning. He is also able to copy geometrical figures, but is unable to copy letters and figures.

In the *fifth* degree, or what Kussmaul calls *literal* agraphia, the patient is unable to write a single letter, and, *a fortiori*, he cannot write a word. In such cases even geometrical figures are only imperfectly copied.

Although the sensory mechanism of speech is not much affected in pure motor aphasia, there can be no doubt that the disorder of the expressive faculty does interfere to some slight extent with the full exercise of the apperceptive faculty. Some persons, especially those who are only half-educated, when reading a written or printed page, pronounce each word half-aloud, or at any rate the most important words in each sentence, and when the power of articulation is lost to these persons the capacity of comprehending what is read is considerably impaired. A clergyman once said to me, "I can only study when the pen is in my hand;" and it is manifest that had this gentleman become the subject of a motor agraphia, his power of comprehending a difficult subject would have been much lessened.

Patients affected with motor aphasia are able to frown, laugh, and smile as usual, and can also play whist and other games requiring skill. In many cases of aphemia the patient is able to sing simple airs, but generally without words, although on rare occasions persons have been known to sing with the accompanying words a few lines of a favourite song, while quite unable to utter any other word. The chief disabilities from which the subjects of motor aphasia labour are brought together and represented in the following table :—

APHEMIA.

THE FACULTIES OF SPOKEN SPEECH.	VARIETIES.				
	1st deg.	2nd deg.	3rd deg.	4th deg.	5th deg.
1. Spontaneous vocal speech in sentences..............	Impaired.	Lost.	Lost.	Lost.	Lost.
2. Repetition of words and reading aloud	Retained.	Retained.	Lost.	Lost.	Lost.
3. A few intelligent replies to questions in single words	Retained.	Retained.	Retained.	Lost.	Lost.
4. Occasional and recurring utterances of no speech value	Retained.	Retained.	Retained.	Retained.	Lost.
5. Grunting sounds, and syllabic utterances, not forming any word	Retained.	Retained.	Retained.	Retained.	Impaired

MOTOR AGRAPHIA.

THE FACULTIES OF WRITTEN SPEECH.	VARIETIES.				
	1st deg.	2nd deg.	3rd deg.	4th deg.	5th deg.
1. Spontaneous writing in sentences..................	Impaired.	Lost.	Lost.	Lost.	Lost.
2. Writing to dictation and copying sentences	Retained.	Retained.	Lost.	Lost.	Lost.
3. Writing and copying imperfectly a few single words	Retained.	Retained.	Retained.	Lost.	Lost.
4. Writing letters of the alphabet	Retained.	Retained.	Retained.	Retained.	Lost.
5. Copying geometrical figures	Retained.	Retained.	Retained.	Retained.	Impaired

Sensory aphasia is much more difficult to analyse than the motor disorders of speech. Leaving the compound forms of sensory aphasia out of consideration for the present, the simple varieties of the affection may be divided into (1) the aphasia of recollection; (2) psychical blindness; and (3) psychical deafness. The reason for selecting these words will appear in the sequel.

(1) In the *aphasia of recollection* the idea of an object, property, or event is represented in consciousness, but it fails to revive the corresponding word in memory; or a person, object, or action may be actually perceived without the corresponding name being recalled to memory. Proper and general names are the words which are most usually forgotten, and the reason for this is not far to seek. As Kussmaul[1] remarks "the more concrete the idea the more readily is the word to designate it forgotten, when the memory fails." We not only can, but we constantly do, think of a friend, mentally see his face, hear his voice, observe his attitudes and gestures, without even recalling his name. We may even think of his particular good and kind actions in the absence of verbal signs, but we can hardly contemplate "goodness" in an abstract sense without the corresponding word being revived in memory. To state the case more generally, it is possible to think about concrete objects by means of mental images without names, as we may suppose to be done in a rudimentary manner by the lower animals, but abstract thought is impossible in the absence of verbal signs.

The aphasia of recollection, or verbal amnesia presents many degrees. Almost every person who has reached middle age, and even most young persons who have suffered from overwork, anxiety, loss of sleep, or from any cause which leads to imperfect nourishment and exhaustion of the brain, must have experienced the *slighter* degree of this affection. The most common form of the disorder is when the subject of it meets a person with whom he is well acquainted but is unable to recall his name. In the slighter degrees the correct name emerges in consciousness after the lapse of a short time, or a wrong name somewhat similar in sound with the right one, such as Stewart for Stead, rises in the memory. The subject is often aware that the name in his mind is not correct, but the wrong one occupies his attention to the exclusion of the other, or it is only after he is corrected, or ceases to think of the name of his friend altogether, that the right one emerges in consciousness.

In the *second* degree of verbal amnesia, the patient is not only

[1] Kussmaul (A.) art. "Disturbances of speech." Ziemssen's "Cyclopædia," 1871, Vol. XIV., p. 759.

unable to recall the name of his wife and children, probably not even his own, but he is also unable to name most if not all of the objects by which he is surrounded, including his bodily organs. The case of R. B. (Case 8) is a good example of this form of the affection. Some patients are quite unable to name most objects presented to them, others name correctly common objects, but a comparatively long time elapses before the name is recalled to memory, while still others give a wrong name, this last condition being subsequently described under the name of paraphasia. In the first degree of verbal amnesia the power of abstract thought is not at all impaired; and, indeed, it is a matter of common observation that those who contemplate the profounder problems of philosophy are what is called absent-minded, and often the subjects of the slighter degrees of verbal amnesia. In the second degree of the affection, however, there can be no doubt that all the faculties of the mind, including the power of abstract thought, are profoundly injured, and that the idea of things, and the laws which govern them, have, in a great measure, faded from the memory, as well as the words by which these are represented. In such cases, however, when a name is either spoken in the hearing of the patient, or presented to him in writing or printed type, he immediately recognises whether or not it is the appropriate designation of a particular object.

Several other varieties of amnesia have been observed, but inasmuch as in these the memory for ideas fails in corresponding, or even in higher degree than that for words, they can hardly be regarded as examples of a special speech disorder, and therefore do not belong to the subject of aphasia. These varieties of amnesia will, consequently, not be discussed at any great length in this place. The verbal amnesia which is so common in middle age increases with advancing years, and in old age it is not unusual to find that the ideas fade from memory even in greater degree than the words by which ideas are expressed. It not unfrequently happens that old people completely forget recent occurrences, while events which occurred in their childhood remain fresh in their memory. In senile amnesia the recent marriage of a friend may be completely forgotten, while the words necessary to describe such an event may be retained and used to describe a similar event which occurred fifty years previously. A partial amnesia of this kind is liable to occur after severe cranial injuries.[1] It is not uncommon to find that a patient after a fall on the head has completely forgotten not only all the events which have occurred for several weeks after the injury, but also those that happened

[1] See Azam "Traumatismes Cérébrales," Arch. géner. de méd., Feb. et Mars, 1881; Ferré "L'amnésie traumatique isolée," Thèse de Paris, 1881; and Bell (J.), ou "A form of loss of memor following cranial injuries," Edin., 1883.

several weeks before it, or the facts of a favourite science or other subject may completely fade from the memory. The faculty of speech is not, however, particularly affected ; the patient is quite able to describe adequately other occurrences more or less similar to those which are forgotten, and the appropriate words fail, not from any special damage to the speech mechanism, but only because the events themselves have become obliterated from the memory.

(2) *Psychical-blindness* admits of being divided into several degrees or varieties, and it is indeed because word-blindness constitutes only one of these varieties that psychical-blindness has been selected as the generic term. In recalling to memory the words read in a book some men employ chiefly visual and others chiefly acoustic images. It was said of Macaulay that he could read and remember an article of a review by simply looking at each page as he turned over the leaves. His mind took in the meaning of the text at a glance, and he could afterwards remember the relative positions of the words on the printed page. Persons who are gifted in a high degree with the faculty of thinking by visual images have the power of remembering with marvellous accuracy a large number of simultaneous facts, such as the position, form, and colour of the various objects in a landscape, the traits which characterise each man in a large company, and the direction and relative position of the streets, with the details of the architecture of the public buildings, of a large city. There can be little doubt that Macaulay owed his intellectual eminence in great measure to the high development in him of the power of thinking by visual images, and if disease had destroyed this power, even without injuring perceptibly any other part of the structure of his mind, he would have been reduced to the level of a very ordinary person. Now in the *first* or slighter degree of psychical-blindness this is what actually happens. The case of a gentleman is recorded by Charcot[1], who had in a high degree the power of picturing in his mind the minutest details of the physiognomy of absent friends, the plan of each town he visited, and the least important features of the scenery through which he passed ; but who suddenly discovered that he had lost this power during a time of great mental anxiety and sleeplessness. Before the attack he could add up several columns of figures at a glance, but in order to obtain the sum of a column of figures after the attack, he had to look at each figure in succession, and to pronounce them in a medium voice ; and even in ordinary reading, each word had to be accompanied by slight movements of the lips and other muscles of articulation. The visual memory of

---

[1] Charcot. "Un cas de suppression brusque et isolée de la vision mentale des signes et des objets (formes et couleurs)."—*Le Progrès Médical*, Tome XI., 1883. Paris, 1883, p. 568.

forms and of colours had almost completely disappeared subsequent to the attack ; he was quite unable to recall the plan of the town in which he lived, the faces of his wife and children, or even the traits of his own person, and he was known to apologise to a supposed individual in a public gallery for barring the way, that turned out to be his own image reflected in a mirror. This gentleman was highly educated, and was perfect master of German, Spanish, and French, as well as of classical Latin and Greek, and although he could still read fairly well in all of these languages, a certain degree of word-blindness was observable in some of them, notably in the German and Greek. The patient was myopic, and his colour vision was slightly defective, but there was no restriction of the field of vision, and no other defect of sight.

The *second* degree of psychical blindness is the disorder already described under the name of *word-blindness*. In this affection the patient is unable to read a single word either from a printed page or in writing; but in most cases he recovers after a time so far as to recognise his own name, the name of the town in which he lives, and a few other simple and very familiar words. In aggravated cases the patient is not only unable to read a single word, but he is also incapable of recognising a single letter, a condition named *literal-blindness*. He can, however, still distinguish geometrical figures and recognise portraits. A person suffering from word-blindness may re-acquire the power of reading without recovering from his visual disability. It is well known that a person who is absolutely blind may be taught to read through the nerves of tactile sensibility by the device of raised letters, and one who remains word-blind may teach himself to read through the nerves of the muscular sense. In a case of word-blindness described by Westphal[1] the patient, at first unable to read his own writing, acquired the power of deciphering it by passing the tip of his index finger over each letter as if he were writing the word a second time. The case of a very intelligent gentleman is recorded by Charcot[2], who was suddenly attacked with complete word-blindness; after a time he taught himself to decipher written or printed words with considerable rapidity by moving the index finger of his right hand in the air as if he were writing. When his right hand was fixed at his back he moved the tip of the index finger over the nail of his thumb and thus deciphered the word.

A *third* degree of psychical blindness may be named *partial perceptive blindness*. This disorder is best illustrated by the condition to which

[1] See Kussmaul, *op. cit.*, p. 776.

[2] Charcot. "Des différentes formes de l'aphasie." "De la cécité verbale."—*Le Progrès Médical*, Tome XI., 1883, p. 441.

Goltz reduced dogs by washing out with a jet of water large portions of the cortex of the brain. The animals operated upon manifested a peculiar imperfection of vision; they could still see and avoid obstacles, but failed to recognise that some of the objects by which they were surrounded were endowed with special properties which constituted them food. The mental condition to which these dogs were reduced was named by Munk psychical blindness, but as we have already said, it will be most convenient to use this term in a generic sense. The foundation of every perception by sight is to be found in the connection between certain lights and shadows falling upon the eye and the feeling of resistance. Rudimentary sight may be regarded, according to Mr. Herbert Spencer, as anticipatory touch. Suppose a man to be seated at a table, and certain lights and shadows fall upon his eyes, which he knows to be cast by a loaf of bread. The knowledge that the sensation of certain lights on the eye is cast by a loaf implies the further knowledge that on stretching out his hand a certain distance he can grasp the object, and that it will offer a certain degree of resistance to his hand. This is the primary or fundamental part of the perception, but it is only a very small part of the ideas called up in men of ordinary intelligence by the sight of a loaf of bread. Every person knows that the loaf can be cut into slices by a knife, and that when a portion of it is taken into the mouth and chewed, it has an agreeable taste, and that it nourishes the body when swallowed. The more educated a man is the larger will be the number of ideas which cluster round a perception. The sight of a loaf, for instance, may revive in the memory of some men the idea of the mill in which the grain was converted into flour, the wheat from which the grain was derived, and the fields in which the wheat was grown; while in persons who have had a special training in chemistry, the sight of it may revive a large mass of knowledge about the properties of gluten and starch, and the physiology of digestion. Most of my readers will know to their sorrow the extreme facility with which the knowledge of the chemical properties of gluten and starch fade from the memory even in the absence of any disease, and a mental condition may be induced by injuries of the head or severe disease of the brain, in which a visual sensation fails to arouse in the mind any of the properties of the object except the fundamental one of its inertia, and if this, the most profound of all relations, viz., that between the sight of an object and the idea of its resistance, is once broken, a person would be regarded as having lost either his sight or his reasoning faculties. If, for example, a person were seen moving his hand over a table as if the light reflected by a loaf were a shadow instead of indicating a resisting substance; or if he were seen walking in an apartment without taking any notice of the shadows cast by objects, and without arresting his movements until he came into collision with the objects themselves, it would be concluded that he was either completely blind or mad. The mental condition to which Goltz reduced the dogs on which he operated was one in which the fundamental relation between the sight of an object and the idea of its resistance was retained, but in which the sight of an object failed to arouse the idea of any of its remaining properties, and consequently such dogs were unable to make any practical distinction between the leg of a table and a leg of mutton, both objects being regarded as obstacles to be avoided. A more or less similar mental condition to that of these dogs has also been observed in man. A peculiar defect of sight was observed by Fürstner[1] in cases of general paralysis of the insane in which the occipital lobes were particularly involved in the disease. The patients could still see large objects and avoid obstacles, but they failed to count small things correctly, or to appreciate the special properties of objects. A coin, for example, when placed in the hand, was recognised as a piece of metal, but the knowledge of the special characteristics which constituted it into a money tender was lost. Partial perceptive blindness can hardly be regarded as an aphasia, but it is mentioned here because it forms the connecting link between true word-blindness and the results obtained in animals by experiments on the sensory cortical centres.

(3) *Psychical-deafness* is used here, like psychical-blindness, as a generic term including various disorders in the intellectual appreciation of words, while the faculty of hearing sounds is not necessarily interfered with. We have already seen that in the recollection of language some persons chiefly employ visual and others acoustic images. It is well known that some men, for instance, have a marvellous faculty of recognising, at first sight, persons whom they had casually met some years previously, whilst others recognise their long-absent acquaintances most readily on hearing them speak. Persons in whom the acoustic word-imagery is highly developed recall to memory the voices of their friends rather than their physiognomy and gestures, the singing of birds and the roaring of cataracts rather than the form and colouring of the objects in the scenery through which they pass, and the hum and bustle of a large town rather than the style of its architecture and the plan of its streets, while they are likely to be more appreciative of music than of painting and sculpture. In most persons, however, both the acoustic and visual imagery are sufficiently developed to find the loss of either faculty a great deprivation. A case of loss of the acoustic word-

[1] Fürstner. "Ueber eine eigenthümliche Sehstörung bei Paralytikern."—*Archiv für Psychiatrie*, Bd. VIII., 1887, p. 162.

78

imagery has neither been observed by ourselves nor have we met with one recorded in medical literature, but we have no doubt that cases of the kind will soon be discovered when a careful search is made, and in the meantime we will name this acoustic disability the *first* degree of psychical-deafness, and give it a place in our subdivisions of this disorder.

Word-deafness constitutes the *second* degree of psychical-deafness. In this condition the patient is not only unable to name correctly any object presented to him, but the names of objects and persons uttered in his hearing also fail to revive in his memory corresponding ideas. The degree in which spoken words fail to call up appropriate ideas differs greatly in different cases. In some cases the patient replies readily to most simple questions so long as the names of objects are avoided, while in other cases the disorder is so profound that the patient fails to comprehend almost every question addressed to him, however simple it may be. In some cases of word-deafness the patient can both comprehend reading and read aloud quite well, but in others the auditory disability is accompanied by a considerable degree of word-blindness. The word-blindness is in some of these cases accompanied by bilateral hemianopsia, and such must be regarded as a compound variety of the two affections; but in other cases bilateral hemianopsia is absent, and in these the word-blindness must be looked upon as being secondary to the word-deafness. That word-deafness should carry with it a certain degree of word-blindness is not difficult to understand.[1] In the acquirement of language spoken speech takes precedence of written speech. An association is first formed between the spoken word and its corresponding perception or idea, and when subsequently the word comes to be seen in writing the association is formed not directly between the written word and the idea, but indirectly through the spoken word. A half educated person can only comprehend what he reads when the words as uttered by himself fall upon his ear, and it is manifest that in such a person word-deafness would entail almost complete word-blindness. And even educated persons, when reading a difficult subject, are much aided in comprehending it by muttering half aloud the principal words of each sentence; and in such persons also word-deafness would cause some difficulty in understanding written language, although not complete word-blindness. The degree of deafness to words varies greatly in different cases. A case of fracture of the brain is reported by Oré,[2] in which the patient failed to understand questions addressed to him in French,

[1] See De Watteville (Dr. A.) "Note sur la cécité verbale." *Le Progrès Médical*, 21, Mars, 1885.
[2] Oré. *Bull. Acad. de Méd.*, 2 S., Tome VII., 1878, p. 1183.

but replied at once when the question was asked in patois. In other cases a still less degree of disability is present. The case of a Russian gentleman is recorded by Charcot,[1] who, besides his native tongue, was master of French and German. After a cerebral attack he failed to comprehend questions asked him in German, but understood quite well when addressed in Russian or French. In some cases the patient understands many questions so long as concrete nouns are avoided, while in other cases the patient fails to give evidence that he understands any word uttered in his hearing. In some few cases the deafness extends even to letters and figures, constituting *literal* deafness corresponding with the literal blindness already described. The perception of music is sometimes preserved even in complete word-deafness, but in most cases the appreciation of music is much impaired. It is, however, abundantly proved that word-deafness does not entail deafness to ordinary sounds; a slight knocking at the door, for instance, being readily heard and its significance quickly interpreted.

Partial-perceptive deafness is not so well known as the corresponding visual affection. The dogs operated upon by Goltz took no notice of calls to which they readily responded prior to the injury, yet they were not deaf, inasmuch as they were observed to start at noises. Now the fundamental relation in an auditory perception is to be found in the association between certain noises and the feeling of danger, and of other sounds and something, such as prey or food, to be enjoyed, and of these the former is probably the more profound of the two. If, for example, a loud or sudden noise be made close to the ear of a person and it fail to cause a start, it would be concluded that the subject was absolutely deaf; but if a person, who affords the usual indications of hearing sounds of moderate intensity, fail to understand the significance of a loud tapping at the door of his apartment, it would be concluded that he is the subject of the partial-perceptive deafness under consideration. Examples of this form of partial deafness are doubtless to be found in our asylums, although the condition has not yet, so far as we know, been adequately described.

In motor aphasia the apperceptive faculty of speech is, as we have seen, not much affected, but the converse of this proposition is very far from being true, inasmuch as in the various forms of sensory aphasia the expressive faculty of speech is the subject of a profound disorder. The different derangements in articulate expression which are observed in sensory aphasia may, following Kussmaul, be grouped together under the term " paraphasia."

---

[1] See Bernard (Désiré). " De l'aphasie et de ses diverses formes." *Thèse de Paris*, 1885, *p.* 154.

Paraphasia presents several degrees according to the intensity of the sensory disorder, and to the nature of the apperceptive faculty which is injured. The *first* or *slightest* degree of the affection hardly oversteps physiological limits, and is liable to occur in persons exhausted from fatigue or hunger, or who are excitable or nervous. A person absorbed in thought, or whose brain is exhausted from some cause, will often find that on meeting a friend a wrong name starts in his memory to the exclusion of the right one. In many cases the wrong name is connected with the right one by similarity of sound, like the example already given of Stewart for Stead, or the familiar Brown for the less familiar Barrett, while in other cases the connection is formed by some more hidden link of association. At other times mistakes are made in grammar and diction, or there may be a transposition of the syllables of two words in the same sentence, a condition which has been named by Kussmaul "syllable-stumbling," but which is doubtless a variety of paraphasia. As an example of syllable-stumbling Kussmaul mentions the case of an absent-minded professor who, before his class, spoke of "the two great chemists Mitschich and Liderlich, meaning Liebig and Mitscherlich." Those who studied medicine in the University of Aberdeen some years ago, will remember that one of our most respected professors was constantly in the habit of committing ludicrous mistakes of this kind, "cus porcuscles" instead of "pus corpuscles," being an example of which we have been recently reminded. It may here be remarked that great hesitation in recalling the correct names of objects and events is of itself a sign of the presence of some degree of sensory aphasia, just as slowness in the execution of certain movements, such as beating the devil's tattoo on the table, is a sign of motor paralysis.

The *second* degree of paraphasia is met with in the second degree of verbal amnesia. The patient being unable to recall to memory the names of objects, is compelled to substitute for the correct word a paraphrase of it or a wholly inappropriate term. A person, for example, who forgets the name of "scissors," may say "the thing you cut with," or of a "chair" "that you sit on." When a wrong name is substituted for the right one, the former is generally connected with the latter by similarity in sound or meaning, or by frequent association in the use of the objects represented by the words. The patient may, for example, say "butter" instead of "mother," "pencil" instead of "pen," or "pepper" instead of "salt."

The *third* degree of paraphasia is observed in those cases of word deafness in which the patient is not only unable to give an appropriate name to any object, but also fails to appreciate the right name when it

81

is uttered in his hearing. In cases of this kind the patient often employs one word, which by long use and frequent association had become automatic in him before the attack, as the designation of every object presented to him. James Lee (Case 5), for example, named every object "public-house." His index finger was "the first public-house," and the remaining fingers, "the second," "third," and "fourth public-house" in succession. His thumb was "the public-house there," and his head "this public-house," at the same time pointing to it. The workhouse, in which he was on several occasions an inmate, was "the public-house up yonder," and so on. At other times the patient may substitute for the appropriate name an unintelligible sound instead of a word. Joseph Lander (Case 4) for example, named a pencil a "punt," and although he knew that this was incorrect every fresh endeavour to name the object ended in it being called a "punt."

A *fourth* and last degree of paraphasia is met with in certain cases of gibberish aphasia. It is very probable that in some cases of this kind of aphasia the derangement of the expressive faculty of speech is caused by a direct injury of the motor mechanism. In the case of John Hand-forth (Case 9), for example, word-deafness was present in a marked degree, but inasmuch as the patient was the subject of a persistent hemiplegia, it is probable that the motor mechanism of speech was like-wise injured. A too exclusive consideration of this case led us to assert that gibberish aphasia is always associated with persistent hemiplegia, but Dr. Clifford Allbutt, whose opinion on such a question is so well deserving of confidence, has, in a private communication, kindly pointed out that this association is not invariable. A further examination of the subject has convinced us that Dr. Allbutt is right. The case of a man, aged 60 years, is reported by Broadbent,[1] who, after a fit of some kind, suffered from paresis of the right side of the face, but no hemiplegia. His speech was mere inarticulate jargon, while he was almost com-pletely word-blind and word-deaf. At the autopsy a large focus of softening was found in the left hemisphere of the brain in the area of the parietal and parieto-sphenoidal branches of the Sylvian artery. In this case the motor mechanism of speech was uninjured, and yet vocal expression was reduced to mere jargon. It is evident, therefore, that the gibberish resulted not directly from injury of the motor, but indi-rectly from injury of the sensory mechanism of speech, and it must consequently be regarded as a paraphasia. The disorder may, therefore, in such cases be named *syllabic paraphasia*. In a very remarkable case

---

[1] Broadbent. "On the cerebral mechanism of speech and thought." *Medico-Chirurgical Transactions*, Vol. LV., 1872, p. 162 *et seq.*

reported by Dr. Osborne[1] the patient, after an apoplectic attack which left no trace of paralysis, gave utterance to a rapid succession of unintelligible sounds when he attempted to speak, while his efforts at reading aloud ended in mere jargon. The patient, however, is said to have understood all that was said to him and also what was written, and was even able to express his thoughts, in writing, only using incorrect words rarely. In this patient the apperceptive faculty of speech appears to have been intact, and his speech disability must consequently be regarded as a variety of aphemia, which may be named *syllabic aphemia*, and corresponds to our fifth degree of aphemia. The case shows that even the gibberish aphasia which results from injury of the motor mechanism of speech may occur in the absence of hemiplegia.

The power of *repeating* words is lost in cases of complete word-deafness, but it is preserved in uncomplicated cases of word-blindness, as, for example, in the case of Robert Marshall (Case 2).

The subjects of sensory aphasia make mistakes in writing, which correspond more or less with those made in speaking, the affection being named by Kussmaul "paragraphia."

Paragraphia admits, like paraphasia, of being divided into several varieties, according to the degree of the affection. The *first* degree of paragraphia occurs in most healthy persons when the brain is exhausted from any cause. It shows itself sometimes by errors in spelling simple words, and at other times by elision, or transposition of words or of whole phrases in a sentence.

The *second* degree is met with in those cases of verbal amnesia in which the patient is unable to recall the names of persons and objects. The patient being unable to recollect most of his concrete nouns, he cannot write them, and he is consequently unable to write continuously anything beyond a few simple sentences, although he can write readily to dictation, especially if the words are frequently repeated in his hearing.

The *third* degree of paragraphia is met with in word-deafness. In this form of aphasia the patient is almost wholly incapable of completing a sentence in writing. The letter written by James Lee (Case 5) to one of his fellow patients is a good example of this degree of the affection.

A *fourth* degree may possibly be constituted by some at least of those cases in which the patient writes whole lines of letters without being able to form a single intelligible word. The case of a woman is mentioned by Hughlings-Jackson[2] who, after a slight attack of aphasia with

[1] Osborne. *The Dublin Journal of Med. and Chemical Science*, 1834, Vol. IV., p. 157, quoted by Bastian, *Brit. & For. Medico-Chir. Review*, Vol. XLIII., 1869, p. 229.

[2] Jackson (Hughlings), *London Hospital Reports*, Vol. I., p. 432, quoted by Bastian (Charlton), *British and Foreign Medico-Chirurgical Review*, Vol. XLIII., Jan., 1869 p. 232.

transient hemiplegia, on attempting to write her name, wrote, "Sunnil, Siclaa, Satreni," and gave her address as "Suncsr Nut Ts Mer Tinn-Lain." It is, however, probable that, as the patient is said to have been very intelligent, the disorder in this case was due to injury of the motor mechanism. The case of a man was reported by myself,[1] who, after an injury to the head, named objects indifferently "two," "tooth," and "measures," and understood but imperfectly what was said to him, while his attempts at reading aloud ended in little better than gibberish. This man fell from the top of an omnibus, and the right side of his forehead was the first part to strike the ground, so that the angular gyrus and the posterior part of the first and second temporosphenodial convolutions of the left hemisphere would be the parts that would suffer from the *contre-coup;* and, as there was no paralysis, the motor mechanism of speech was not likely to have been injured. Both the symptoms and the nature of the injury therefore pointed to the conclusion that his speech disability was a sensory aphasia. On being asked to write his name—William Abson—he took hold of the pen with the air of one accustomed to its use, and in a bold hand, but with con-siderable hesitation, wrote, "Wuagageng Abreaghrer." As he wrote each letter he named one aloud, but the written and spoken letter never corresponded with one another, and very few of either corresponded with the letters of the name he was attempting to write. From this case it may be concluded that a sensory aphasia may give rise to so great a disorder in writing that every syllable and almost every letter of the writing fails to correspond with the word attempted to be written, and the disorder may consequently be named *syllabic or literal paragraphia.* Other disorders emerge when the subject of a sensory aphasia attempts to write to *dictation.* In cases of verbal amnesia patients can write correctly to dictation, but in aggravated cases the words have to be frequently repeated to them. The word-deaf, of course, are unable to write to dictation. A few days ago James Lee (Case 5) was asked to write down "This is a fine day," and, after taking some time for con-sideration, he wrote "whiskey," which at present has replaced "public-house," and "glass of beer" as his favourite name for all objects. The word-blind can, if educated, write to dictation, but they write better with their eyes closed than open, and are unable to read what they have written. Robert Marshall (Case 2) could write a few words to dicta-tion, but if he raised his pen while writing a polysyllabic word, he got confused and was afterwards unable to complete it.

The power of *copying* written or printed words is also deranged in

---

[1] Ross (J.) "A case of amnesic aphasia occasioned by a fall on the head." *The Lancet,* Vol. II. 1881, p. 904.

many cases of sensory aphasia, while it is preserved in others. James Lee (Case 5), who was word-deaf, and could not write a word to dictation, could copy correctly in a beautiful running hand from a written or printed page. Robert Marshall (Case 2), who was word-blind and could write to dictation, was unable to copy a single letter of a written or printed word. Some patients suffering from word-blindness can copy words very slowly, taking letter by letter, and forming it as it is in the copy, whether written or printed, just as if it were a geometrical figure. In other cases of sensory aphasia a still greater success in copying is attained; but it still continues to be very imperfect. William Abson, the case just alluded to, who was word-deaf, and, most likely, word-blind, on being asked to copy "With deep feeling," wrote with great deliberation, and with frequent glances at the copy, "Weeth deap flneearer."

Some of the subjects of sensory aphasia commit mistakes on attempting to read aloud, this condition being named by Kussmaul "paralexia."

Paralexia admits also of being divided into several varieties, according to its degree of intensity. The *first* degree of paralexia occurs in absent-minded and nervous people, and consists of slurring over some words altogether, transposing others, and similar ludicrous mistakes. In the profound verbal amnesia, in which the names of most objects are lost, the patient is able to read aloud and generally correctly, or only makes slight mistakes like an absent-minded person; but he forgets what he has read immediately afterwards.

In the *second* degree mistakes are so frequently committed that the sense of what is read is almost completely lost to the listener, although it is possible that the patient himself may understand the meaning of what he is reading. Considering the extreme degree of the speech-disorder from which James Lee (Case 5) suffered, he read aloud with wonderful correctness; but he hardly ever read a sentence without mispronouncing some of the words, and thus marring the sense to the listener.

In a *third* degree of paralexia every word, or almost every word, is represented by syllabic sounds to which no definite meaning can be attached, as in cases of gibberish aphasia. The disorder observed in Robert Marshall (Case 2), in which the patient composed his own reading, but not a word of which corresponded with any in the text, is doubtless allied to the profounder degrees of paralexia, although it is somewhat difficult to classify.

In partial perceptive blindness the patient must commit serious mistakes in conduct, a condition which has been named by Kussmaul "apraxia." A patient, for example, who fails to recognise that a piece of metal he holds in his hands is a money tender of a certain specified value must necessarily commit gross mistakes in purchasing commodities. The case of a young man is reported by Gogol who, after an injury to the head, suffered from a complicated form of aphasia, manifestly including word-blindness. Kussmaul,[1] who quotes the case, says of this patient, "he urinated in the wash-basin, bit a morsel out of a cake of soap, and did other things of a like character, which proved that he confounded material objects with one another. He performed preposterous acts, or, in other words, suffered from apraxia." Corresponding mistakes in conduct may doubtless be committed by those suffering from word-deafness, but an example of the kind is not known to me.

Paramimia is a word used by Kussmaul to designate the disorders of intellectual gesture, which may occur in sensory aphasia. A patient may, for example, nod assent when he wishes to give a negative answer, or *vice versa*. Those disorders of gesture are, however, not well ascertained, and consequently they need not detain us longer.

The chief disorders, both of the apperceptive and expressive faculties of speech which occur in sensory aphasia, are brought together in the following table :—

[1] Kussmaul. *Op. cit.*, p. 797.

SENSORY APHASIA.

THE FACULTIES OF SPEECH	VERBAL AMNESIA.		PSYCHICAL BLINDNESS.			PSYCHICAL DEAFNESS.		
	1st degree.	2nd degree.	Loss of Visual Mental Representations.	Word-Blindness.	Partial Perceptive Blindness.	Loss of Acoustic Mental Representations.	Word-Deafness.	Partial Perceptive Deafness.
I.—The Apperceptive Faculties.								
1. Perceptions and ideas revive in memory corresponding verbal signs	Impaired.	Lost.	Retained.	Retained or impaired.	Impaired.	Retained.	Lost.	Lost.
2. Perceptions and ideas revive in memory visual representations of objects	Retained.	Retained.	Lost.	Lost.	Lost.	Retained.	Retained.	Retained.
3. Graphic verbal signs revive in memory corresponding ideas	Retained.	Retained.	Retained.	Lost.	Lost.	Retained.	Retained.	Retained.
4. Visual perceptions revive in memory the special properties of objects	Retained.	Retained.	Retained.	Impaired.	Lost.	Retained.	Retained.	Retained.
5. Perceptions and ideas revive in memory the acoustic representations of sounds	Retained.	Retained.	Retained.	Retained.	Retained.	Lost.	Lost.	Lost.
6. Acoustic verbal signs revive in memory corresponding ideas	Retained.	Retained.	Retained.	Retained.	Retained.	Retained.	Lost.	Lost.
7. Acoustic perceptions revive in memory the special properties of objects	Retained.	Retained.	Retained.	Retained.	Retained.	Retained.	Retained.	Lost.
II.—The Expressive Faculties.								
1. Spontaneous vocal speech	1st degree paraphasia.	2nd degree paraphasia.	Retained.	Retained.	Impaired.	3rd degree paraphasia.	4th degree paraphasia.	4th degree paraphasia.
2. Repetition of words	Retained.	Retained.	Retained.	Retained.	Retained.	Impaired.	Lost.	Lost.
3. Reading aloud	Retained.	Retained.	Retained.	Lost.	Lost.	Retained.	Impaired.	Impaired.
4. Spontaneous written speech	1st degree paragraphia.	2nd degree paragraphia.	Retained.	Impaired.	Impaired.	Retained.	3rd degree paragraphia.	4th degree paragraphia.
5. Writing to dictation	Retained.	Impaired.	Retained.	Impaired.	Impaired.	Impaired.	Lost.	Lost.
6. Copying	Retained.	Retained.	Impaired.	Lost.	Lost.	Retained.	Retained.	Retained.
7. Conduct	Retained.	Retained.	Retained.	Impaired.	Apraxia.	Retained.	Impaired.	Apraxia.

Having subjected the chief symptoms of aphasia to analysis, it now remains for us to connect the clinical varieties of the affection with disease of certain parts of the nervous mechanisms of speech. Attempts have been made, from time to time, by different authors to accomplish this end by means of a diagram. The first attempt of the kind was made by Baginsky,[1] and he was soon followed by Wernicke[2] and Spamer,[3] but those schemes, being somewhat crude, need not detain us, and we shall at once proceed to describe briefly the very important diagram constructed by Kussmaul.[4] In Figure 6, J represents the ideational centre or centre of conceptions; B, the acoustic sensory centre, and B′ the visual sensory centre for words. C is the motor centre for the co-ordination of the movements required for spoken speech,

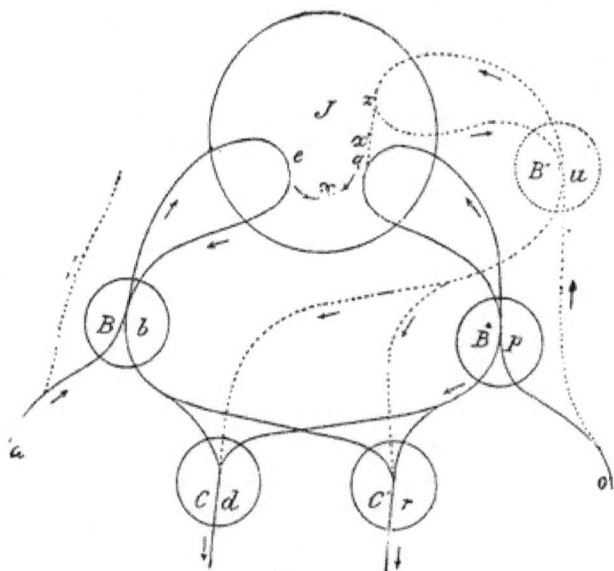

Fig. 6.

and C′ the corresponding centre for written speech ; a is the acoustic, and o the optic nerve ; a b c b d is the collective acoustic track for spoken speech, and o p q p r the corresponding optic track for written speech ; a b d is the track for the imitative speech of children and o p r is the track for the copying of uncomprehended words.

[1] Baginsky (D.). "Aphasie in Folge schwerer Nierenkrankungen.—Urämie." *Berliner klin. Wochenschr.*, Bd. VIII., 1871, p. 441.

[2] Wernicke (C.). "Der Aphasische Symptomen-complex." Breslau, 1874.

[3] Spamer (C.). "Ueber Aphasie und Asymbolie nebst Versuch einer Theorie der Sprachbildung," *Arch. f. Psychiatrie*, Bd. VI., 1875, p. 507.

[4] Kussmaul (A.). "Disturbances of speech." *Ziemssen's Cyclopædia of the Practice of Medicine*, Vol. XIV., 1878, p. 779.

c b d is the track for giving utterance to thoughts, and *q p r* for writing down thoughts.

c x q represents the connection between acoustic and visual word images, and by means of it the translation of written into spoken words is rendered possible, and *vice versâ.*

b r and *p d* represent respectively the tracks between the centre for spoken images and the motor centre for writing, and that between the centre for written images, and the motor centre for spoken speech.

a b r is the track for writing down an uncomprehended spoken word, and *o p d* the one for reading aloud an uncomprehended written word.

a b c b r is the path for writing down dictated words that are understood, and *o p q p d* the one for reading aloud comprehended written words. The dotted lines indicate paths from the nerves of sense passing to the ideational centre through other centres than those of acoustic and visual word images; such, for example, as a tract through the optic nerve leading to a centre for the appreciation of gestures and pantomimic speech generally, and through the acoustic nerve to a centre for the appreciation of music. The arrows indicate the direction of the conduction.

Such, then, being Kussmaul's diagram of the nervous mechanisms of language, let us see how he explains the various speech disorders by disease of particular paths and centres.

The *deaf-mute* is incapable of using the tracks *a b c b d* and *a b c b r*, but the centres C′ and C can be reached through *o*, and thus the power of writing and speaking can be acquired. In writing he can copy without understanding the text through the path *o p r*, and with understanding of it through the path *o p q p r*. In learning to speak, the acoustic centre B for spoken words cannot be reached, and consequently a visual centre B″ for the appreciation of the movements of the lips and the other parts of the articulatory apparatus must be developed to take its place. When the patient mimics the movements of articulation without comprehension, he uses the track *o u d*, and with comprehension *o u z u d*. By the use of the central connection *z x′ q*, it is possible to translate mimic word-images into written word-images, and *vice versâ.*

In *complete motor aphasia* the centres C and C′ are injured, and the paths passing through them cannot be used. In aphemia the centre C, and in motor agraphia the centre C′, is alone injured.

In those cases in which the patient has lost the faculty of spontaneous speech while retaining the power of reading aloud, the path *b d* is injured, while the centre C with the path *p d* are unaffected. In the *aphasia of recollection* the path *c b* is temporarily blocked, and the patient is unable to name objects spontaneously. When, however, the sound of the name falls on *a*, the impulses passing through the path *a b d* enable the patient to utter the word in a reflex manner, and the extra stimulus caused by the word suffices to re-establish the arc *b c b*. The extra stimulus necessary to open the path *c b* may also be supplied through *o p q x*, by means of the written name.

In *word-deafness* the centre B, and in *word-blindness* the centre B′ is injured, and the tracks leading to and from these centres are injured in varying degrees. In *word-deafness with paraphasia* the track *a b c* is injured, while the track *c b* still permits conduction to the more or less disordered centre B, and the track *b d* is intact. In *word-blindness*, with retention of the ability to *copy* writing, but without comprehension, the track *p q p* is obstructed, while the track *o p r* is uninjured. In *word-blindness*, with retention of the ability to write spontaneously and from dictation, the track *p q* is obstructed, while the tracks *q p r* and *a b r* are preserved. The case of word-blindness recorded by Westphal, in which the patient was enabled to read writing by passing his finger over each letter, seems, in Kussmaul's opinion, to prove that there is not only a track leading from *b* to *r*, but also from *r* to *b*. At this stage of our enquiry it is scarcely necessary to add that this assumption is superfluous, inasmuch as the knowledge is acquired through the muscular sense.

Another very important diagram has been constructed by Charcot,[1] and although his plan is similar in its main outlines to Kussmaul's scheme, he has, with his usual illustrative power, rendered it more striking to the senses, and more easily comprehended at a glance than any hitherto devised. The author believes that the word *bell* (cloche), with its corresponding object, is particularly well adapted for the study of the development of language and its subsequent disorganisation by disease. The object is shown at the lower part of the diagram, and its sound is represented by a wavy line passing to the left, and upwards through the auditory nerve to the acoustic centre C A C for common sounds. After a time the sound of the word *bell* (cloche) falls on the ear, and is conducted to the special centre for spoken words C A M, which lies near to, or is part of, the common acoustic centre. By frequent repetition an inseparable association is formed between these two sounds, so that the presence of the one revives the idea of the other in consciousness, an operation which is represented structurally by the union of the paths of the separate sounds in the common centre of conceptions or the ideational centre I C. The subject is now in possession of two means by which the idea of the object may be revived in memory, namely, the sound of the bell itself, or the sound of the

[1] See Bernard (Désiré). "De l'aphasie et de ses diverses formes." *Thèse de Paris*, 1885, p. 45.

word. But in order that the individual may be enabled to awaken in others the idea of the object, it is necessary that a new centre should be developed—the centre C L A—for the regulation of the movements of the organs of the articulation of words. If the impulse to action comes to the motor centre of articulation C L A directly through the path marked 1 from the sensory centre C A M, then the word is merely an imitative sound, without meaning to the person uttering it; but if it comes from the ideational centre I C, through the conducting path 2, then the word is uttered with full comprehension of its meaning. On

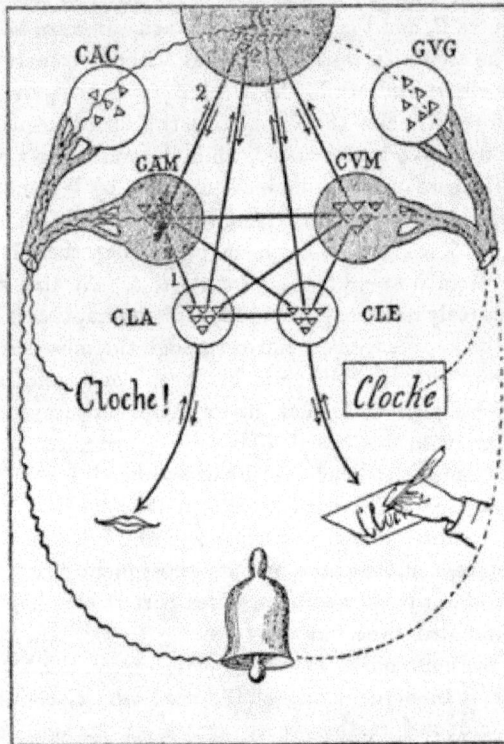

CAC   GVG   2   CAM   CVM   CLA   CLE   Cloche!   Cloche

the visual side of the development of language, the object and the printed word *cloche* give rise to impulses represented by the dotted lines, which, on passing through the optic nerves, make a first impression on the common visual centre C V C and the special visual centre for words C V M respectively. The organisation in these centres, caused by the two separate visual impressions, becomes united after a time in the ideational centre I C, and this in its turn becomes connected with the motor centre for the regulation of the movements of the hand in writing, C L E. The arrows indicate the direction of the conduction, and it will be observed that on the lines which connect the mouth and the hand with the centres for the regulation of the movements of articulation C L A, and of writing C L E, respectively, and the lines connecting these centres with the ideational centre I C, the arrows point both upwards and downwards. This is to show that when a person utters or writes a word he becomes conscious of the operation through the nerves of muscular sense, without the aid of the special senses. It is unnecessary to give a further description of this diagram, inasmuch as the reader who has mastered our discussion of Kussmaul's scheme will readily interpret for himself the kind of speech disorder that will be caused by disease of particular centres and conducting paths.

In a very elaborate and able paper on aphasia by Lichtheim, translated into English by de Watteville,[1] several diagrams, or rather

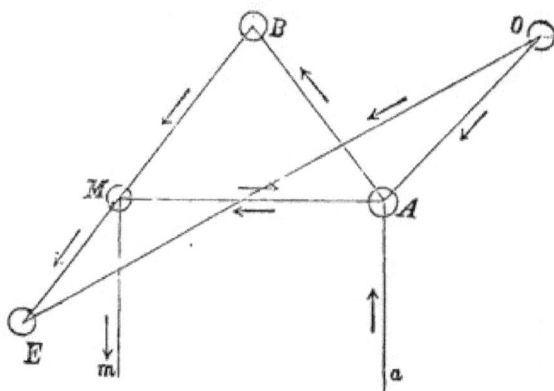

Fig. 7.

several modifications of the same diagram, are given to illustrate the various speech disorders, and of these we select Fig. 7 as being the simplest, if not the most complete. In this diagram B represents the centre for conceptions, or the ideational centre. A is the centre of acoustic, O that of visual, word-images; M is the motor centre for the regulation of the movements of articulation, and E for the regulation of the movements of the hand in writing. The commissural fibres connecting these centres will be readily intelligible without further description. The author believes that the different clinical forms of aphasia may be reduced to seven types, and that each of them is caused by lesion of one of the centres, A and M, or of one of the five conducting paths, A M, B M, M m, A B, and a A.

[1] Lichtheim (Prof. L.). "Ueber Aphasie." Aus der medicinischen Klinik in Bern. *Deutsches Archiv f. klin. Med.*, Bd. XXXVI., Leipzig, 1885, p. 204. De Watteville (Dr. A.) "On aphasia." *Brain*, Vol. VII., Lond., 1885, p. 433.

The annexed table, borrowed from a very excellent abstract of Lichtheim's paper by Thomsen,[1] gives at a glance the author's seven types of aphasia, with the centre or conducting path injured in each :—

Functions of Speech	Paths necessary for the different Functions.	Forms of Aphasia caused by Lesion of the following Centres and Conducting Paths.						
		1 M.	2 A.	3 A M.	4 B M.	5 M m.	6 A B.	7 a A.
Volitional speech	B M m.	Lost.	Retained.	Paraphasia.	Lost.	Lost.	Paraphasia.	Retained.
Volitional writing	B M E.	Lost.	Retained.	Paragraphia or lost.	Lost.	Retained.	Paragraphia.	Retained.
Reading aloud	O A M m.	Lost.	Lost.	Paraphasia.	Retained.	Lost.	Retained.*	Retained.
Repetition of words	a A M m.	Lost.	Lost.	Paraphasia.	Retained.	Lost.	Retained.*	Lost.
Writing to dictation	a A M E.	Lost.	Lost.	Paragraphia.	Retained.	Retained.	Retained.*	Lost.
Understanding of spoken words	a A B.	Retained.	Lost.	Retained.	Retained.	Retained.	Lost.	Lost.
Understanding of written words	O A B.	Retained.	Lost.	Retained.	Retained.	Retained.	Lost.	Retained.
Faculty of copying	O E.	Retained.	Retained.	Retained.	Retained.	Retained.	Retained.	Retained.

* Retained, but without comprehension.

A careful study of this table will show that the clinical analysis of

[1] Thomsen. "Ueber Aphasie."—*Centralblatt fur klinische Medicin*, Bd. VI., Leipzig, 1885, p. 417.

cases is carried out only to a very imperfect degree. It will be at once seen that the first and fourth types are only different degrees of aphemia in association with motor agraphia, while the fifth represents aphemia without motor agraphia. The seventh type represents a moderate degree of word-deafness, while the second and sixth represent word-deafness and word-blindness variously combined. The third type now remains for consideration. In the example which Lichtheim gives, the patient on being asked "'What was there for supper?' answered, 'Bread, meat, potatoes,' with only two mistakes."[1] On being asked to relate his history he strung together numerous words, of which only one now and then could be made out, such as "evening, five and twenty and." The patient could understand spoken and written language, and could also repeat correctly short words uttered in his hearing, but in the repetition of words generally and in loud reading he committed the same kind of mistakes as in voluntary speech. This case is doubtless allied to our first degree of aphemia, a condition in which the patient comprehends both spoken and written speech, and is able to utter a great many words, but cannot complete a sentence. The fact, however, that at the autopsy the lesion was found chiefly in the area of distribution of the posterior branches of the Sylvian artery, shows that the sensory mechanism of speech was injured. It will suffice at present to say that Lichtheim's third type corresponds with the disorder of speech which Dr. Broadbent has named "inability to express the relations between things," and which he believes to be caused by disease of an assumed propositionising centre. Such cases are believed by Wernicke[2] to be caused by injury of the fibres which connect the sensory with the motor centres of speech, and he has consequently named them *commissural aphasia*. But we shall return to the consideration of these cases after giving a brief account of Dr. Broadbent's views of the sensory disorders of speech.

Dr. Broadbent[3] sets out in his explanation of speech disorders with the proposition that all muscular movements are performed under the direction of a "guiding sensation." If, for example, the palm of the hand of a person asleep be tickled, impulses are conducted inwards to a particular level of the spinal cord, and thence are reflected outwards to the muscles which close the hand. When, however, the individual is

---

[1] Lichtheim, *Brain, loc. cit.*, p. 445.

[2] See Wernicke (C), *Lehrbuch der Gehirnkrankheiten für Aerzte und Studirende*, Bd. II., 1881, p. 205.

[3] Broadbent (W. H.) "A case of peculiar affection of speech with commentary." *Brain*, Vol. I., London, 1879, page 484 *et seq.*; see also, "On a case of amnesia, with post-mortem examination." *Medico-Chirurgical Transactions*, Vol. LXI., 1878, p. 147; ' On a particular form of amnesia; loss of nouns." *Ibid*, Vol. LXVII., 1884, p. 249.

awake, the outgoing portion of the reflex arc can be utilised by the cortex of the brain, and then voluntary closure of the hand takes place. The nuclei of the motor fibres of the peripheral nerves in the spinal cord are, therefore, subservient both to centripetal impulses coming from the periphery, and to centrifugal impulses from the cortex of the brain. But the centrifugal impulses from the cortex are initiated and controlled by centripetal impulses going from the periphery to the cortex. It thus appears that each movement is represented in the anterior grey horns of the cord by a group of connected cells, and that this group may be called into activity either by centripetal impulses going from the periphery to that level of the spinal cord in which these cells are situated, or by centrifugal impulses descending to this level from a higher nerve centre. In Dr. Broadbent's words, " a motor cell-group is formed under the guidance of a sensory cell-group on the same level, and when formed is made use of by a higher centre." In the case of speech the "motor cell-group" must combine into orderly action—the thoracic muscles to obtain an expiratory current of air, the laryngeal muscles for phonation, and the muscles of the lips and tongue for articulation. This motor cell-group Dr. Broadbent names for convenience the *word group*, and he believes that it is organised in the corpus striatum. When the cells of the word-group are called into activity by centripetal impulses on the same level the action is reflex, and the resulting contractions simply represent a complicated muscular adjustment without any reference to intellectual expression; but on the activity of the group being evoked from the cortex, then the movement becomes subservient to speech. The cortical outlet for speech is situated in the third left frontal convolution, and the cortical guiding sensory centre for spoken language is situated in the superior temporo-sphenoidal convolution (auditory centre). In accordance with the annexed diagram, lesion of S (the motor centre for speech) will cause motor aphasia, while lesion of A (the auditory perception centre), or of *a s* (the fibres which connect the inlets and outlets) will cause different forms of sensory aphasia. A hypothetical explanation is thus afforded for three disorders of speech. In lesion of S, the "way out" for all the muscular adjustments concerned in intellectual expression is destroyed. In lesion of *a s*, the line of communication between the guiding sensory centre and the motor outlet is damaged, and mistakes in words recognisable by the patient occur; while in lesion of the sensory centre A, mistakes in words occur of which the speaker remains unconscious. But still higher centres than those named are brought into use in intellectual expression, and disease of these produces various complicated disorders of speech. "The formation of an idea of any external object," says Dr. Broadbent,[1] " is the combination of the evidence respecting it received through all the senses; for the employment of this idea in intel'ectual operations it must be associated with and symbolised by a name. The structural arrangement corresponding to this process I have supposed to consist in the convergence from all the 'perceptive centres'

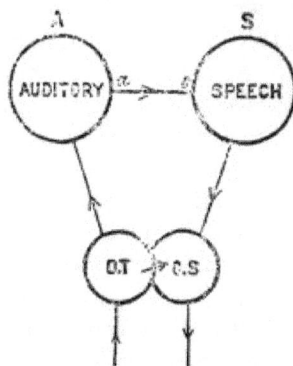

of tracts of fibres to a convolutional area (not identified), which may be called the 'idea centre' or 'naming centre.' This will be on the sensory, afferent, or upward side of the nervous system; its correlative motor centre will be the propositionising centre, in which names or nouns are set in a framework of other words for outward expression, and in

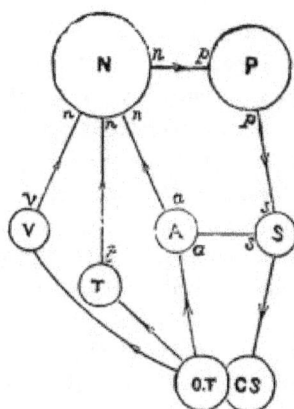

which a proposition is realised in consciousness or mentally rehearsed. If we are to have a seat of the faculty of language, it would be here rather than in the third left frontal convolution, with which, however, it may possibly be in close proximity. Expressing this by a diagram, we

[1] Broadbent (W. H.) *Brain*, Vol. I., 1879, p. 494.

have V, A, and T, the visual (angular gyrus, Ferrier), auditory (infra-marginal Sylvian gyrus), and tactual (uncinate gyrus), perceptive centres, sending converging tracts of fibres, *v n*, *a n*, *t n*, to N, the 'naming centre.' Here the perceptions, from V and T (smell and taste are omitted for the sake of simplicity), are combined with an idea, which idea is symbolised by the name reaching N through A, which has always, in the experience of the individual, been associated with the object. P is the propositionising centre, in which the phrase is formed, its relations with N and S being sufficiently clear."

According to this scheme, lesion of the naming centre N would cause loss of the memory of names or nouns, leaving the patient able to express himself imperfectly in words indicative of relations and attributes. Lesion of P, the propositionising centre, would render the patient unable to construct a sentence, although retaining the use of names. This condition Dr. Broadbent[1] illustrates by the case of a patient, who, in endeavouring to explain that he had two brothers in America, said—"Brother — New York — brother — 'Merica — letter — two brothers in America—letter." The patient could understand spoken and written language, but his attempts at reading aloud ended in gibberish. He was unable to write to dictation, and, with the exception of his own name and the names of his brothers—which he wrote quite well—he was also unable to write spontaneously. He could copy short words quickly and correctly, but a long word he took down slowly and in schoolboy characters, and as he wrote each letter he named it aloud, *and always wrongly*. This form of speech disorder Dr. Broadbent regards as an "inability to express the relations between things"; it corresponds with Lichtheim's third type, and with Wernicke's commissural aphasia, but we reserve further consideration of it at present.

Lesion of the visual perceptive centre V, or of its channel of communication *v n* with the naming centre N, would explain cases of word-blindness, while lesion of the auditory perceptive centre A, or of its channel of communication *a n* with the naming centre would explain word-deafness.

A very good diagram illustrative of the disorders of speech is given by Professor Grainger Stewart,[2] and although I believe that he postulates too many centres both on the ingoing and on the outgoing side of the speech mechanism, his scheme, taken in conjunction with his remarks on aphasia, is well worthy of careful study. No good, however, would result from dwelling at greater length on this part of our subject.

[1] Broadbent (W. H.), *Brain*, Vol. I., 1879, p. 486.

[2] Stewart (T. Grainger). An introduction to the study of the diseases of the nervous system." Edin. 1884, page 181 *et seq.*

Instead of entering upon a criticism of the diagrams figured and described in the foregoing pages, we prefer to indicate briefly our own views of the nature of the connection existing between the various speech disorders and the lesions which underlie them. In the first place we shall deal with motor aphasia, because it is more simple, and consequently more easily handled, than the sensory forms of the affection.

The first and most important proposition which we will endeavour to establish is that motor aphasia is the result of a genuine paralysis. Some pathologists have offered much opposition to this view. They assert, and with truth, that the subject of a pure motor aphasia is often capable of moving his lips freely, of protruding his tongue and employing it in the act of deglutition, and of using his vocal apparatus in singing, without giving evidence that the muscles engaged in these actions are suffering from any degree of paralysis. They also point to the well-ascertained fact that in cases of pure motor agraphia all the other movements of the hand may be effected with ease, and without the gross strength of the grasp being diminished. Those who urge these objections believe aphemia to be caused by a want of motor co-ordination, and they have consequently named the affection *ataxic aphasia*. That aphemia is due to a motor inco-ordination we do not deny; so is diplopia, but in most cases it is paralytic also. In paresis or paralysis of the sixth nerve, for instance, the patient is unable to direct the axes of vision to one object, and he consequently sees two objects instead of one, and so far there is an inco-ordination of the ocular muscular movements, but it is caused by paralysis of one of the muscles of the eyeball. And to call aphemia by the name of *ataxic aphasia* is only less absurd than it would be to call cases of double vision from paralysis of one or more of the ocular muscles by the name of *ataxic diplopia*, because the paralysis is not always so easily discovered in the former as in the latter affection. The brain is, indeed, so far as its motor functions are concerned, an organ by means of which the simple movements which are regulated by the spinal cord are variously combined and co-ordinated so that complex bodily adjustments are produced, and when the cerebral motor organisation is wholly destroyed by disease the co-ordination is altogether lost, or when it is partially destroyed the co-ordination is so disordered that the balanced muscular contractions which are requisite for effecting delicate movements fail to take place. In every case of partial paralysis from cerebral disease, therefore, there is a want of motor co-ordination; but this motor disorder is in no way comparable with the genuine ataxia of *tabes dorsalis*, in which not a trace of paralysis is to be discovered.

In order to illustrate the theory that aphemia is a paralytic disorder, let us attend to the course of the development of the

G

movements of the hand. In the infant at birth the hand closes on anything that touches lightly the palm, this movement being altogether reflex and regulated by the spinal cord. When the infant is a little older he begins to use the hand as an organ of prehension, and the acquired movements, which are now voluntary and regulated from the brain, become more and more complicated by use, each complication of movement being represented by a new complication of structure in the cerebrum. Let us now suppose that the faculty of writing is not developed until after the infant has reached adult age and has acquired the full strength of manhood. This supposition is made in order to show that the delicate muscular adjustments which have to be acquired in learning to write have added nothing to the strength of the hand as an ordinary prehensile organ, and it is easily conceivable that disease may destroy the new organisation in the brain which is the structural correlative of those delicate movements, without diminishing in the slightest degree the strength of the hand for ordinary purposes. Now this is what occurs in pure cases of motor agraphia. In most cases loss of the faculty of writing is accompanied by impairment of the power of the hand as an organ of prehension, and it is then said that the agraphia is accompanied by more or less of hemiplegia, but it is important to recognise that motor agraphia is itself an indication of a slight degree of hemiplegia, even in those rare cases in which the gross strength of the hand is not perceptibly diminished. Similar remarks apply to aphemia ; it is of itself an indication of hemiplegia even in the rare cases in which it is unaccompanied by any perceptible paralysis of the face or tongue.

If further proof be required of the paralytic nature of aphemia and motor agraphia, it is found in the fact that these speech disorders are accompanied, except in the rarest instances, by some degree of hemiplegia. And again, in experiments on animals, electrical excitation of the posterior part of the third frontal convolution, which is the seat of the lesion in aphemia, gives rise to opening of the mouth, with protrusion and retraction of the tongue, thus showing the essential identity of function of this part of the brain with the other motor centres of the cortex. The phenomena which attend unilateral convulsions teach the same lesson. It is now well recognised that an attack of unilateral epilepsy is followed by some degree of paralysis of the parts which were the seat of the spasm. When the spasm begins in the foot, the leg, and when it begins in the hand, the upper extremity, are found to be more or less paralysed for some time after the cessation of the spasm ; and when the spasm begins at the angle of the mouth, the attack, when right-sided, is not only followed by a considerable loss of expression on the side of the face that was the subject of spasm, but also by a decided aphemia. From these considerations it may be concluded that aphemia and motor agraphia are a paralysis of highly specialised movements ; that in rare cases the special movements of articulation, and of the hand in writing, are paralysed without the general movements of the respective organs being interfered with ; but that in most cases not only the special but the general movements also suffer to some extent, and then only is it popularly recognised that the speech disorder is accompanied by, or is a part of, hemiplegia. Before proceeding further, therefore, it is desirable for us to study somewhat minutely the mechanism by which hemiplegia is produced, and the phenomena which attend it, for it is in this way alone that we can hope to throw light upon several obscure problems connected with aphasia and motor agraphia, which still remain unconsidered.

It may be stated as a general proposition that the muscles of one lateral half of the body are regulated from the cerebral hemisphere of the opposite side. The centres for the regulation of the movements of the opposite side of the body are situated on the outer convex surface of the cerebral hemisphere—the so-called motor area of the cortex. These centres are connected with the anterior grey horn of the spinal cord of the opposite side, and to a less extent with that of the same side by the long fibres of the pyramidal tract. Now hemiplegia of the opposite side of the body may be caused by extensive disease of the motor centres themselves, but the ordinary form of hemiplegia is usually caused by compression or rupture of the fibres of the pyramidal tract, as they come together to form one bundle between the basal ganglia in the internal capsule. It is found, however, that interruption of conduction through the whole of the fibres of the pyramidal tract of one hemisphere does not paralyse all the muscles of the opposite side of the body. In profound hemiplegia some of the muscles are completely paralysed, while others are scarcely, if at all, affected. It was first pointed out by Dr. Broadbent[1] that the muscles which remain comparatively unaffected by paralysis in hemiplegia are those which, like the muscles of respiration, are associated in their actions with the corresponding muscles of the opposite side. In cases of severe hemiplegia the muscles of the hand, the actions of which are independent of the muscles of the other hand, remain persistently paralysed, while the muscles of respiration, which always act in conjunction with the corresponding muscles of the opposite side, may show a slight degree of feebleness at first, but recover completely in a few days. The explanation offered by Dr. Broadbent for the fact that the muscles which are bilaterally associated in their actions

[1] Broadbent (W. H.), *British and Foreign Medico-Chir. Review*, 1866, p. 477.

have a relative immunity from paralysis is that these muscles are inner-
vated from both cerebral hemispheres, and consequently when the motor
centres or conducting paths of the one hemisphere are injured, the
muscles in question receive their impulses to action from the uninjured
hemisphere. He believes that the connection between the motor nervous
mechanisms of the two sides of the body is effected by means of commissural fibres in the medulla oblongata and spinal cord.

In the annexed diagram B and B′ represent the cortices of the right
and left hemispheres of the brain, M M′ represent motor centres, and
N and N′ nerve nuclei in the medulla oblongata or spinal cord. P and P′
represent the conducting paths in the pyramidal tracts which form a
crossed connection between the cortical motor centres and the spinal
nuclei, while m and m′ represent the nerves which pass out to the
muscles. 1, 2, 3, 4, and 5 represent lesions in different situations. When

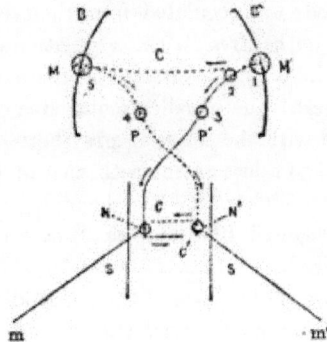

the muscles of the one side of the body, as those of the hand, act inde-
pendently of the corresponding muscles of the opposite side, the above
centres and conducting paths constitute the whole of the motor
mechanism, and when a lesion at 1 injures the motor centre, or at 3 the
conducting path, the muscles on the opposite side become permanently
paralysed. When, however, the muscles are bilaterally associated in
their actions, another mechanism is superadded. It consists of commissural fibres connecting the nerve nuclei—the one (c) conducting from
right to left and the other (c′) from left to right—and now when a
lesion 1 or 3 injures the motor centre or the conducting path in the one
hemisphere, impulses pass from the uninjured motor centre M of the
other hemisphere to the nerve nucleus N′ of the opposite side, and then
across and through the commissure c′ to the other nerve nucleus N, and
to the muscles, through the motor nerves m and m′. In this manner
the muscles of both sides receive impulses to action from the one
hemisphere B. The principle which has just been described has been

happily named by Dr. Broadbent the "bilateral association of the nerve nuclei of muscles bilaterally associated in their actions."

The laryngeal muscles belong to the group of muscles which are bilaterally associated in their actions, and they escape almost entirely in ordinary hemiplegia; and although the other muscles of articulation, such as those of the mouth and tongue, are implicated in the paralysis, yet they are not profoundly affected, and the loss of power which they undergo only gives rise to slight articulatory difficulties. In left-sided hemiplegia, for instance, in which the special mechanism of speech is not usually affected, the voice of the patient remains almost entirely unaltered after the attack, while the paralysis of the facial and lingual muscles which is present only gives rise to some thickness of speech and a difficulty in pronouncing polysyllabic words. When hemiplegia occurs in infancy, before the development of speech, and in the lower animals, the muscles of the vocal apparatus are not, any more than those of respiration, affected by the occurrence of paralysis on either side, but subsequently to the evolution of speech in man the conditions are entirely changed. To the simple nervous mechanism which we have just been considering, a highly complicated structure has become gradually developed, and by means of it the simple movements of the lips, tongue, and vocal apparatus which are alone possible in early infancy and in the lower animals, are now combined into the complicated movements of articulate speech. The chief characteristic of this superadded mechanism is that it is unilateral. It has already been seen that speech disorders are caused, except in rare cases, by disease of the left hemisphere of the brain, and in the annexed diagram the special motor centre M′ for speech is situated in the left hemisphere B′, while the corresponding centre M in the right hemisphere regulates the general actions, which the articulatory muscles would have been capable of performing even if the faculty of speech had never become developed. It is scarcely necessary to observe that in the speech centre M′ both the general and special movements of the articulatory muscles are organised, although we shall find it convenient sometimes to call it the *special* motor centre of articulation in opposition to the centre M of the other hemisphere, which may be named the *general* motor centre. In the diagram the special motor centre M′ is connected by a continuous line P′ with the nerve nucleus N of the opposite side, and through the commissural fibre c′ with the nerve nucleus of the same side, so that the muscles of both sides of the articulatory apparatus are regulated from a centre in one hemisphere. The general motor centre M in the right hemisphere is also connected with the nerve nuclei of both sides, as is represented by the dotted lines P and C, but as already remarked the complex combinations of contractions requisite to articulatory speech cannot be effected by it.

From the foregoing remarks it will be at once apparent that complete destruction of the special motor centre M′ will abolish the faculty of the expressive faculty of speech, but the general movements of the articulatory muscles on both sides will still be effected by means of the centre M. From this statement, it almost follows as a corollary, that partial damage of the special articulatory motor centre will give rise to a partial disorder of the expressive faculty of speech. Now in all partial lesions of nerve centres, the latest evolved structure is the most liable to injury. The inroads of disease conform, both as regards structure and function, to the law of dissolution ; the mode of invasion being from the complex to the simple, and from the special to the general. On examining the structure of the cortex of the brain, it will be seen that it is composed of several super-imposed layers of cells ; and that the cells near the surface are destitute of processes, and therefore do not form any definite connection with one another, while the cells of the internal layers have numerous processes, by means of which they become connected with one another, and some of them even with the motor ganglion cells of the medulla oblongata and spinal cord. Now the nervous impulses regulating actions, which are frequently repeated in the experience of the individual and of the race, tend to pass more and more with each repetition through the caudate cells of the cortex, and these being, as we have seen, definitely connected with one another, the nervous impulses pass through them without meeting much resistance, and the less the resistance offered to the passage of the impulses, the less is the resulting action attended by consciousness. Such actions are sometimes called automatic and at other times reflex actions of the brain, but they really are psychical actions which are effected in an unconscious manner. But the impulses which regulate unaccustomed actions must pass through unused channels, or in other words, must pass through the round cells near the surface of the cortex which have not formed definite connections with one another. In passing through these cells, the impulses meet with a considerable degree of resistance, and much of the force generated is expended, not in effecting the desired action, but in producing a new organisation in the round cells of the cortex, and this is the process which is the correlation of consciousness. It follows, therefore, that a superficial lesion of the cortex of the brain will abolish in a motor centre the power of performing unaccustomed actions, while leaving intact the power of performing the actions which have been so frequently repeated that they have become automatic, or are effected in an unconscious manner. The application of this theory to the explanation of the motor disorders of speech is manifest. A superficial lesion of Broca's convolution may destroy the power of uttering all unaccustomed words, while leaving intact the power of uttering a considerable number of the words which have been early acquired and have been so frequently repeated that their utterance has become automatic. A lesion passing somewhat deeper in the cortex may entirely abolish the power of giving utterance to spontaneous speech while leaving unaffected the automatic power of repeating words uttered in the hearing of the patient. A still further degree of injury may arrest the power of repeating words, while leaving intact the power manifested by infants in their first endeavours at speech of uttering a large number of vocables or syllabic sounds, amongst which no word having a definite meaning can be detected, this forming one variety of gibberish aphasia. It is very probable that the power of giving expression to "recurring utterances" and oaths, which is so frequently left to persons who are otherwise the subjects of a complete aphasia, is, as Dr. Hughlings-Jackson conjectures, organised in the general articulatory motor centre, and may, therefore, be preserved in cases in which the special centre is thoroughly disorganised.

But if destruction of the special motor centre M' in the annexed diagram gives rise to aphemia, it may be expected that the same disorder of speech will be caused by a lesion (3) which ruptures the motor conducting path, and thus severs the spinal nuclei from the centre. Now right-sided hemiplegia is almost always at first accompanied by some degree of aphasia,[1] but when the lesion is situated, as it usually is, in the lenticular nucleus, and compresses or ruptures the motor conducting path in the internal capsule, the power of speech is recovered sometimes in a few days, and at other times in a few weeks or months. When, however, the speech centre itself is disorganised, the speech disorder remains more or less permanent in old people, while in young persons the power of speech is only slowly acquired, most probably by a new organisation being formed in the motor centre of the opposite hemisphere. How, then, can we explain the persistency of the speech disorder in the cases in which the centre itself is diseased, as compared with the relatively speedy recovery which occurs when the conducting path is alone injured? The most common form of hemiplegia is that which is caused by hæmorrhage into the lenticular nucleus. The speech motor conducting path is now proved to pass in the anterior segment of the internal capsule, and in many cases of hæmorrhage into the lenticular nucleus this portion of the capsule remains almost uninjured, while the conducting paths for the limbs which pass in the anterior part of the posterior segment are

[1] See Broadbent, "Cases of Spurious Aphasia."—*Transactions of the Clinical Society*, Vol. XI., 1878, p. 37.

considerably damaged. Immediately after the occurrence of the hæmorrhage conduction through the speech motor path is impaired partly by the shock of the injury and partly by the pressure of the clot, and consequently aphemia results; but as the effects of the shock pass away, and the fluid portion of the effused blood is absorbed the conductivity of the fibres is restored, and the speech disorder rapidly disappears. But even in the cases in which the anterior segment of the capsule is seriously damaged, recovery from the speech disorder appears to be more rapid than in the cases in which the centre is destroyed. The difference between the results of injury to the centre and the conducting path may be illustrated by the electric telegraph. Suppose that the central office in Manchester were blown up by dynamite, all communication from the centre of the city to other towns would be interrupted until a new structure of some kind was formed; but if the conducting wires from Manchester to London were alone injured by a snow storm communication would be immediately restored, as messages could be sent from Manchester to Liverpool, and thence to London. Now, Dr. Broadbent believes that something of this kind occurs in the aphasia which is caused by injury of the motor conducting path. The corpus callosum is still believed by most anatomists to be a great hemispherical commissure, although the recent observations of Professor Hamilton, of Aberdeen, appear to cast some doubt upon the theory, and its fibres are supposed to connect symmetrical parts in the two hemispheres. It has been suggested by Dr. Broadbent that when the speech motor conducting path P' in the annexed figure is interrupted by lesion at 3, the impulses from the special speech centre M' find their way through the fibres C of the corpus callosum to the corresponding motor centre M of the opposite hemisphere, and thence through the conducting path P and the commissure c to the nerve nuclei in the medulla. A process of this kind will occupy some time, because the nervous impulses have to make their way through channels which have been more or less unused, but the time occupied will be much shorter than that required for the formation of a completely new organisation in the general motor centre at M. As an additional argument in favour of this theory, Dr. Broadbent urges the fact that a lesion (2 in the figure), which is situated so high up in the white substance of the hemisphere that it will interrupt both the speech motor conducting path and the fibres of the corpus callosum which connect the posterior parts of the third frontal convolutions, will cause an aphasia which will be as persistent as that resulting from destruction of the centre itself.[1]

[1] See Pitres. "Lésures du centre ovale." Paris (*Obs. XXX.*, vi.), 1877, p. 94 *et seq.*

But if the theory of motor aphasia which has been advanced in these pages be true, destruction of the two motor centres M and M′, or of the two motor conducting paths P and P′, ought to give rise to a complete paralysis of the muscles of articulation. Now this is what actually happens. A large number of cases of this kind have now been reported under the titles of pseudo-bulbar paralysis, labio-glosso-laryngeal paralysis of cerebral origin, and other names, but I can only allude here to one or two examples. Perhaps the most remarkable case hitherto reported in which the centres themselves were destroyed is one recorded by Dr. Barlow.[1] The subject was a boy, aged ten years, who was suffering from aortic regurgitation, and had an attack of right hemiplegia with aphasia, from which he made a good recovery. Four months afterwards he had an attack of left hemiplegia, which was accompanied not only by aphasia but also by paralysis of the muscles of mastication, of those concerned in the first act of deglutition, and of those of articulation. The patient died from the results of the aortic lesion, and at the autopsy the anterior branch of each Sylvian artery was found blocked by an embolus, and a focus of softening, about the size of a shilling, was observed in the cortex of each hemisphere, involving the interior extremity of the ascending frontal and the posterior extremities of the second and third frontal convolutions. Many cases of labio-glosso-laryngeal paralysis, from disease of the motor conducting paths of the cerebrum, are now recorded, and we have ourselves reported several cases of the kind.[2] The first which came under our observation was kindly sent by Dr. Leech. The patient presented all the symptoms commonly observed in cases of progressive labio-glosso-laryngeal paralysis, but at the autopsy, conducted by Prof. W. H. Young, "each cerebral hemisphere presented a single well-defined cystic cavity, containing clear straw-coloured fluid, and occupying the position of the lenticular nuclei," while the medulla oblongata was free from disease. As these cavities, especially the one on the right side, were larger than the normal size of the lenticular nuclei, the fluid doubtless compressed and injured the motor conducting paths of the internal capsules. No more striking proof than these cases afford could well be imagined in favour of the theory, that Broca's convolution is functionally identical with the other motor centres of the cerebral

[1] See Lepine : "Note sur la paralysie glosso-labiée cérébrale en forme pseudo-bulbaire."—Rev. mens. de Med., Tome I., 1877, p. 909. Kirchoff: "Cerebrale glosso-pharyngo-labial-paralyse mit cinseitigem Herd," Arch. f. Psychiat., Bd. XI., 1880, p. 132. Barlow: "On a case of double hemiplegia, with symmetrical lesions."—The British Medical Journal, Vol. II., 1877, p. 103.

[2] See Ross (J) "On labio-glosso pharyngeal paralysis of cerebral origin."—Brain, Vol. V., 1882-3, p. 143; also Raymond (F) et Artaud (G). "Contributions à l'étude des localisation cérébrales."—Arch. de Neurologie, Vol. VII., Mars, 1884, p. 145.

cortex, and of the corollary, which is deducible from it, that the speech disorder—aphemia—caused by destruction of this convolution, is a genuine paralysis, and what is true of aphemia is equally true of motor agraphia.

Motor aphasia is, as we have seen, a paralysis of the special movements of articulation constituting *aphemia*, and of those of the right hand in writing, constituting *motor agraphia*. Some pathologists, however, believe this statement to be imperfect, and they supplement it by saying that this form of aphasia is a loss of the memory of those special movements. Now if this theory of loss of memory be true as applied to the explanation of the inability of a patient suffering from aphemia to execute the articulatory movements necessary to the expression of words, it must be equally true as an explanation of the inability of a hemiplegic patient to execute the movements necessary to normal locomotion. But were we to say that the inability of a patient suffering, for example, from a crural monoplegia from cortical disease, was due to a loss of memory of the movements of locomotion on the affected side, the absurdity of the statement would be immediately recognised, and yet it is not a whit more absurd than the statement that aphemia is due to a loss of memory of the movements of articulation. For our part we agree with Dr. Bastian[1] in thinking that outgoing currents from the cortex to the muscles are free from all subjective accompaniments, and arrest of these currents will consequently not interfere directly with the memory, although it may do so indirectly. The memory of the contractions of the muscles of articulation has doubtless played an important part in the development of the expressive faculty of speech, just as the memory of the contractions of the muscles of the lower limbs has played in the development of the power of locomotion, but the memory in both cases is correlated with the activity of a cortical centre in relation with afferent fibres from the muscles, and therefore the centre is endowed with a sensory and not a motor function. It is possible that the sensory cortical centre of muscular sense may so far coincide with the motor centre for the muscle that the former is constituted by the outer layers of small cells, the activity of which we have already seen to be the correlative of consciousness, while the latter is constituted by the inner layer of caudate cells, the activity of which is not accompanied by any feeling. But if the sensory and motor centres of a muscle or group of muscles are superimposed so as to occupy the same area of the cortex, is it not an excess of refinement to make a distinction between them? Our reply is that nature makes such a distinction. A cerebral paralysis is not always

[1] Bastian (H. Charlton). "The Brain as an Organ of Mind," p. 529; and "Paralysis: Cerebral, Bulbar, and Spinal," London, 1886, p. 107.

accompanied by loss of the muscular sense in the affected muscles, and loss of the muscular sense may be present, as in profound hysterical hemi-anæsthesia, without being accompanied by muscular paralysis. And as loss of the muscular sense in one of the lower extremities produces only a slight disorder of locomotion without entailing paralysis of the limb, so loss of that sense in the muscles of articulation might produce a slight disorder of the expressive faculty of speech, but it certainly would not arrest the articulatory movements to the extent observed in aphemia, and loss of the muscular sense in the muscles of the right upper extremity would not paralyse the special movements of the hand to the extent observed in motor agraphia. We have no hesitation, therefore, in coming to the conclusion that aphemia and motor agraphia are forms of motor paralysis, and that the speech disability present in these affections is not accompanied by loss of memory of words, this inference being fully attested by the fact that the subjects of these disorders understand vocal and written speech, although unable to give expression to their thoughts in words.

Let us now turn our attention to *apperceptive aphasia*. The different forms of speech disorder comprised in this category are caused by disease of the sensory mechanisms of speech, and now that the symptoms of these affections have already been subjected to a detailed analysis, the interpretation of the clinical phenomena in terms of structure need not detain us long. The first assertion which we will venture to make is that as aphemia and motor agraphia are a motor paralysis, so apperceptive aphasia is a sensory paralysis. Considerable difficulty has already been experienced in realising that motor aphasia is a paralytic disorder, and the conception of apperceptive aphasia as a sensory paralysis will not by any means be found to be easier. There is, however, no difficulty in comprehending that complete blindness is a sensory paralysis; and it is also comparatively easy to realise that the profound amblyopia, in which the subject sees the form of an object without being able to appreciate its colour, is a partial sensory paralysis. Now, suppose that a person has his eyes directed to a piece of soap on the table, which is not in any way disguised so as to appear to the healthy eye other than it is, and yet he grasps it and puts it to his mouth as something to be eaten, would it not at once be suspected that that person was partially blind? This action might, of course, result from moral perversity, without any degree of blindness; but defective vision might lead to such an error of judgment as would give rise to it in the absence of any other mental peculiarity. In other words, a certain patch of colour falling upon a defective organ of vision failed to arouse in consciousness certain qualities belonging to the object, while it suggested other qualities which did not belong to it, and thus led to a serious error in judgment. When, therefore, the presentation of a common object to the eye fails to revive in consciousness certain previously well-known qualities belonging to the object, this may be taken as a sign of a partial paralysis of vision. When, for example, the general paralytic patient described by Fürstner failed to recognise that a coin placed in his hand was a money tender, he manifested partial blindness. His conduct, doubtless, also revealed loss of memory and defective judgment, but it did not differ in any essential respect from the conduct of a person who, from having lost the sense of colour owing to advancing white atrophy of the optic discs, mistakes a sovereign for a shilling. In the case of white atrophy, it is at once recognised that the mistaken conduct was the result of partial blindness; but when we pass from the non-revival of the more common properties of matter, such as is met with in cases of achromatopsia, to the non-revival of the less common properties, such as is met with in the disorders already named partial perceptive blindness and deafness, and in word-blindness and word-deafness, we only pass from a general to a special form of partial sensory paralysis. That word-deafness is a partial sensory paralysis may be aptly illustrated by my own mental condition with regard to the French language. I can read French, especially French medical literature, with nearly as much facility as English, but in listening to a conversation in French I am almost absolutely word-deaf. I can detect a few common verbs, pronouns, and adjectives, but hardly ever a noun, and, on the whole, the conversation is to me a series of sounds which blend with one another in inextricable confusion. Another peculiarity is that if I were asked to give the French name of any object presented to me, even including my bodily organs, I would fail, without considerable time for thought, in probably nine trials out of ten; but if the French name were pointed out to me on a printed page, I would instantly identify the object. I am also unable to write French to dictation, but can copy from a written or printed page, and with a full understanding of the meaning of what is written. My condition with regard to the French language is closely analogous to the condition of James Lee (Case 5) with regard to the English, and, indeed, all forms of language. He understands scarcely any word uttered in his hearing, and cannot name correctly any object presented to him; but when he sees the name in written or printed characters, he immediately identifies the object. He can also copy from a written or printed page, and understands the meaning of what he has written, but he is unable to write to dictation anything beyond his name and address. My disabilities with regard to the French language are manifestly due to the fact that my ear has not been educated, or in other words, to the fact that the necessary structure has not been organised in my auditory cortical centre ; while the disabilities of Lee are due to the fact that the structure which had been organised in his auditory cortical centre is now destroyed by disease ; in the one case there is a want of evolution of the necessary organisation, in the other there is a dissolution of a previously acquired structure. Similar remarks apply to word-blindness, although I am unable to illustrate the condition from my own experience so adequately as word-deafness. It appears to me, however, that the condition of Robert Marshall (Case 2), when looking at a printed page of English is, in some respects, similar to my own when looking at a page of Hebrew text. On scanning a page of Hebrew text, I not only cannot decipher a word, or even a letter, but I have even difficulty in imagining how any other person can attach a definite meaning to such a confused assemblage of dots and dashes. It is quite conceivable that I might have acquired the power of thinking and conversing in Hebrew without having learnt to read the language, and had I done so, my mental state would have been closely analogous to that of Marshall, with regard to English. I should then be able to name all objects presented to me, and to converse in Hebrew as Marshall does in English, and like him, I should be unable to read a printed page or to decipher a single letter. I should be able, like some of the subjects of word-blindness, to copy printed words in Hebrew in printed and written in written characters, and without being capable of attaching any meaning to what I had copied, though, unlike most of those who suffer from word-blindness, I should be unable to write to dictation. But notwithstanding the great similarity existing between cases in which there is a dissolution of one of the sensory mechanisms of speech and those in which the mechanism has been imperfectly evolved, yet the former present mental peculiarities which are not paralleled in the latter. For instance, if I were asked to write down the French phrase of "It is a fine day," I should never write anything so totally unlike it as "whiskey," as James Lee (Case 5) did ; and if I were asked to read from a printed page of Hebrew, I should not, even supposing I were able to converse in the language, give utterance, like Robert Marshall (Case 2) to my own composition, fully believing that I was interpreting correctly the text before me. It need hardly be expected, however, that the brutal dissolution of nerve centres by sudden disease, like the softening caused by the occlusion of an artery, will ever exactly parallel in its results a simple arrest of the organisation of the same nerve centres at a particular stage of their evolution. On the whole, then, it may be concluded from physiological considerations, conjoined with the results obtained from post-mortem dissections, that word-blindness and word-deafness are a partial sensory paralysis, resulting from damage of the visual and auditory cortical centres respectively.

We must now proceed to interpret as far as possible those minor cases of apperceptive aphasia which we have named *verbal amnesia*. The slighter degrees of verbal amnesia are met with whenever the brain becomes exhausted from over-work, loss of sleep, or from being imperfectly supplied by a sufficient quantity of healthy blood. In such cases the lesion is a functional one, and is widely distributed over the whole brain, especially over its surface, where the nutritive processes are most active. Manifestly such cases are not well calculated to throw much light on the disorders of the speech mechanisms which underlie aphasia. The severer forms of verbal amnesia are, however, caused most probably by a destructive lesion, and consequently these disorders demand detailed examination. The second degree of verbal amnesia, as already described, consists of those cases in which the patient understands what is said to him, and is able to read, but cannot name any object presented to him, and consequently is unable to use nouns in his conversation. In the case of R. B. (Case 8), the patient was either unable to name the objects presented, or could only name them after a long pause for consideration, but he instantly recognised the correct name when it was uttered in his hearing, or when it was presented to him in writing, and he could at once repeat it. In a case described by Dr. Broadbent,[1] the patient, during the five years in which he was under observation, was unable to name any object presented to him, and although he understood all that was said, and promptly obeyed when told to say "one, two, three," during an examination of the chest, he was unable to repeat any noun uttered in his hearing. It is quite manifest that Dr. Broadbent's case is a degree of apperceptive aphasia intermediate between the speech disorder of R. B. (Case 8) who, although unable to name objects, understood the name, and could repeat it when uttered in his hearing, and that of James Lee (Case 5), who could neither name objects, understand the name when uttered in his hearing, nor repeat it. Dr. Broadbent believes that in his case the communication between the ideational or naming centre and the propositionising centre[2] was destroyed. Before endeavouring to account for the clinical variety of apperceptive aphasia which Dr. Broadbent has named "loss of nouns," we ought to ascertain whether such a disorder of speech has ever been observed, or is possible. Is it true that an individual may lose the use of concrete nouns while the other parts of speech remain unimpaired? We do not believe it. The case described by Dr. Broadbent as "loss of nouns," and other similar cases, merely show that the whole structure of

[1] Broadbent (W. H.). "On a particular form of amnesia. Loss of nouns."—*Medico-Chirurgical Transactions*, Second Series, Vol. XLIX., Lond., 1884, p. 249.

[2] Broadbent (W. H.). *Loc. cit.*, p. 258.

language has been damaged, and in such cases the only words left to the individual are those which have been most frequently used in his experience, and which have, therefore, become most deeply organised in him. In these cases there is a dissolution of language, in which the most special parts are the first, and the most general the last to disappear, and if concrete nouns disappear in greater degree than other parts of speech, it is chiefly because they form a very special part of language. Dr. Broadbent's patient gave expression to such phrases as "I am very glad to see you," "I am very much better to-day, thank you," "Oh! yes, very well indeed," while James Lee's (Case 5) conversation consisted of such statements as "I am quite well, there is nothing the matter with me, if I could only speak it; but I cannot speak it at all." Now, although these phrases are couched in the developed language of civilisation, it is manifest that the thoughts expressed in them are hardly above the level of the first expressions of infancy, or of the thoughts which may be supposed to have been uttered by aboriginal man in his first attempts at articulate speech. There is a widely prevalent belief, even amongst cultivated psychologists, that nouns are the first parts of speech to be developed. This notion is due most probably to the fact that *ma-ma* is the first articulate sound of the infant, and from this it is inferred that the infant's first step in the acquisition of language is to call its mother by name. But in so far as the infant associates the sound *ma* with its mother it is only a particular name, and is more like to a nickname than to our developed conception of the general name, *mother*; and in the earliest stage of the acquirement of speech by the infant, it is doubtful if the exclamation *ma* has even the signification of a particular name. It is probable that it is associated in the infantile mind with the pleasant sensations which contact with the mother, and especially with the breast of the mother, brings along with it, and consequently the signification of the word *ma* in the mind of the infant is more nearly expressed by the imperative mood of the verb "to come" than by our general name "mother." The signals of the social animals teach us a similar lesson. When a rook, for example, utters "Caw! caw! caw!" as a danger signal to his fellows it cannot be supposed that he uses a noun having a meaning corresponding to our abstract word "danger," or to our words for the usual causes of danger, such as man, gun, gunpowder, &c. The nearest interpretation to the signal in our language would be the imperative mood of the verb "to fly." And when language, as a whole, is subjected to analysis, it is found that all words are derived from simple predicative roots, signifying, to go, to come, to divide, to eat, &c., and demonstrative roots such as, this, that, I, he, it, which are simply used to point at an object or person.

The rudiments of the demonstrative pronoun may even be recognised in the signal cry of the rook when taken along with the movements of the one giving the signal, inasmuch as the direction of his own flight indicates their course to the other members of the flock, so that the cry and movements are tantamount to our phrase "fly in this direction." The devices which persons suffering from apperceptive aphasia employ to overcome their disability, remind us of the manner in which aboriginal man used his few scanty words to convey his thoughts. Were James Lee (Case 5) to be presented the object *hay*—the present staple food of cattle—we have no doubt but that he would describe it in the following terms:—"I know it quite well; it is a glass of beer;" and on being remonstrated with for this answer he would add, "I have seen that thousands of times; *they eat that.*" In aboriginal times the staple food for cattle was acorns and beech-nuts, and it would become necessary to have a name for these fruits at a very early period in the development of speech. In the Sanskrit there is a root *bhag*, meaning to divide, and from the act of making a division of the fruits which had been collected for the cattle, the word came to be used also for the act of eating them, as is shown by the Greek word φαγ, to feed, to eat, and, after a time, the same root was used to designate the trees from which these fruits were derived, as in the Latin *fagus*, the Greek φηγός, oak, and the English *beech*.[1] The science of language teaches unmistakably that the language of aboriginal man consisted almost entirely of verbs, demonstrative pronouns, and a few adverbs of time and place, and that the names of even common objects are always derivative, and, consequently, of much later growth[2] than the roots them-

[1] See Max Müller (F.) "On the Origin of Reason." *The Contemporary Review*, Vol. XXXI., 1877-78, p. 490.

[2] In a discussion which followed the reading of a paper on aphasia by Dr. Gairdner before the Philosophical Society of Glasgow, March 7th, 1866, the Rev. H. W. Crosskey is reported to have said, "that naming objects was an exceedingly difficult thing, and involved a process of thought difficult for a child to accomplish; and, therefore, it was found that the verb, of all utterances, was the first name, and that to give a name to an object showed a considerable amount of predication and will, so that the arrest of the faculty of the mind to grasp proper names, instead of being an arrest of a primeval quality of the mind, was an arrest of one of the latest and most elaborate actions of it."—(Gairdner (W. T.). "On the Function of Articulate Speech," p. 39, Glasgow, 1866.) It is to me a matter of regret that Dr. Gairdner's remarkable paper did not come under my notice until after the series of papers, which are here collected in one volume, had already appeared in the *Medical Chronicle*. Its perusal has convinced me that I have not done full justice to the share which my own countrymen have taken in differentiating motor and apperceptive aphasia. The distinction between the two forms of speech disorder is suggested by Dr. Gairdner, as an alternative to some other hypothesis, when he says, "or we must admit *two perfectly distinct* kinds of aphasia, only one of which affects the *ideation* of language, so to speak, while the other affects, in some complicated way, as yet imperfectly studied, but perhaps differing from paralysis properly so-called, the *innervation* of language, or rather of speech, while it leaves the *ideation* of it, on the one hand, and the mechanism of it through the writing hand, on the other, absolutely, or nearly, intact."—(*Op. cit.*, p. 20.) About the same time with Dr. Gairdner, his friend, the late Professor Sanders of Edinburgh, gave expression to a similar view in still more

selves, and it is only what might have been expected that in the
dissolution of language caused by disease, nouns should disappear
from the vocabulary of the patient before the parts of speech which
have been first developed, and, therefore, most deeply organised. It
will thus be seen that, in the cases of apperceptive aphasia in which
the patient is unable to use nouns, the whole structure of language is
profoundly injured, and that the subject of this disability is reduced, so
far as language is concerned, to the condition of the infant or the
remote savage in their early attempts at giving articulate expression to
their thoughts, or in understanding the articulate expressions of others.
This view is, however, very different from Dr. Broadbent's theory that
the patient has lost the use of nouns by the destruction of conducting
paths between a naming and a propositionising centre, while the remaining parts of speech have remained unimpaired.

"Another derangement of speech," says Dr. Broadbent,[1] "is where
names are more or less remembered, but there is loss of the faculty of
constructing a sentence which shall convey the ideas to be expressed
regarding them—that is, of the power of framing a proposition. The
lesion here will be in the propositionising centre." The case by which
Dr. Broadbent illustrates this form of aphasia is one in which the patient,
in endeavouring to express that he had received a letter from a brother
in America, said—"Brother—brother—'Merica—letter—New York—
two brothers in America." The same patient, however, in describing
the mode of onset of his attack, said, "Evening, evening—put down
my cigar—smoking, smoking not a quarter of an hour—all at once
couldn't speak." These broken sentences are a very imperfect statement

definite language. "Two kinds of aphasia," says Professor Sanders, "have accordingly been
distinguished. First, amnesic aphasia, loss of speech depending on a defective memory of words,
and, therefore, to some extent, a psychical defect ; and, second, ataxic aphasia, where the loss of
speech is due to lesion of a supposed cerebral apparatus of co-ordination for the movements of
articulate speech—a defect of nervous mechanism only."—(Edin. Med. Journal, Vol. XL, 1885-6,
p. 813.) But not only did Professor Sanders distinguish these two forms of aphasia clinically, he
even went so far as to suggest a possible difference in the position of the lesion in each disorder.
In the case (Mackie) reported by him, the symptoms of motor (ataxic) and apperceptive (amnesic)
aphasia were combined, and at the autopsy one focus of softening was found in Broca's convolution, and a second near the posterior extremity of the fissure of Sylvius—the superior marginal
convolution. Referring to the second focus, Professor Sanders says, "Probably this additional
lesion may account for the greater loss of memory seen in Mackie than is observed in other
cases."—(Loc. cit., p. 822.) The distinctions between motor and apperceptive aphasia are clearly
recognised by Dr. W. Ogle ("Aphasia and Agraphia," St. George's Hospital Reports Vol. II, 1867,
p. 117), but the next great advance in the appreciation of the differences between these two forms
of speech disorder was made by Dr. Bastian, in an altogether remarkable paper, entitled "On the
Various forms of Loss of Speech in Cerebral Disease."—(British and Foreign Medico-Chir. Review,
Vol. XLIII., 1869, pp 299, 470. See also, Bastian, "The Physiology of Thinking," The Fortnightly
Review, January, 1869; "On the Muscular Sense," British Medical Journal, April, 1869; "The Brain
as an Organ of Mind," Third Edition, p. 601, et seq. ; and "Paralysis: Cerebral, Bulbar, and Spinal,"

[1] Broadbent (W. H.) "A case of peculiar affection of speech, with commentary."—Brain
Vol. I. (Lond. 1879), p. 496.

H

of facts, but they contain several parts of speech besides the names of things. The speech disorder of this patient was ushered in by an attack of right-sided hemiplegia, from which he made a rapid recovery; he is said to have understood both spoken and written speech, and had been able to name some objects, and not others, but his attempts at reading aloud ended in mere jargon. A case of this kind of disorder of speech, so far at least as vocal expression is concerned, was sent to me a few weeks ago by Dr. Harris, but as I had only one short interview with the patient I was unable to take full notes of his symptoms. He was a man of about 45 years of age, and I gathered from him that while on a visit to London, about ten years ago, he was suddenly seized with some kind of fit, and was conveyed to Guy's Hospital, where he was attended by Dr. Taylor. On regaining consciousness he found himself paralysed on the right side of the body and unable to speak, but he insisted on leaving the hospital and came back to Manchester the same night. Sometime afterwards he returned to Guy's Hospital, and remained there for three weeks under the care of Dr. Goodhart. He had almost entirely recovered from his paralysis, but had now given up all hopes of his power of speech being restored, and was undergoing treatment at the Hospital for Diseases of the Chest, not with the view of obtaining a cure for his "tongue," but for an attack of pleurisy, from

---

1886, p. 103, *et seq.*) The subject is approached by Dr. Bastian from the double standpoints of the physiology of thinking and clinical analysis, and his main thesis is that "words are revived in the cerebral hemispheres as *remembered sounds*," in opposition to the hypothesis advanced by Bain, and adopted by Hughlings-Jackson, that the material of our recollection is a *suppressed articulation*. If the latter hypothesis were true, both the amnesic and so-called ataxic disorders of speech would necessarily be caused by lesion of the motor mechanism of speech; while, if the former be accepted, the amnesic disorders will be caused, as we now know they are caused, by lesion of the cortical cerebral centres. Dr. Bastian's theory is tacitly adopted throughout these pages; it has, indeed, become so much the common property of psychologists that we are apt to forget to whom we owe its first enunciation, or rather, its first application to the explanation of the phenomena of aphasia. The theory itself had already, as is acknowledged by Dr. Bastian, been briefly, but very distinctly, stated by Mr. Herbert Spencer.—("Principles of Psychology," 1870, Vol. I., p. 187.) The next great advance in the differentiation of motor and sensory aphasia was made by Dr. Broadbent, in a paper entitled, "On the Cerebral Mechanism of Speech and Thought."—(*Medico-Chirurgical Transactions*, Vol. LV., 1872, p. 162.) In this paper the author reported a case of what is now recognised clinically as word-blindness, and in which, after the death of the patient, the lesion was found in the angular gyrus, and neighbouring convolutions. In this contribution, as well as in all his subsequent writings, Dr. Broadbent sees clearly that the disorders of speech which had hitherto been described under the name of amnesic aphasia, are caused by lesion in the area of the posterior branches of the Sylvian artery, and that they result from injury of the sensory mechanism. In the text the recognition of apperceptive or sensory aphasia is attributed chiefly to Wernicke; but, without wishing to detract from the great value of his work, it is only right to point out that it did not appear until 1874, and, therefore, subsequently to the writings which have just been quoted. It is unnecessary for me to speak of the immense strides which the hypothesis of amnesic aphasia, as a disorder of the sensory mechanism of speech, has made since Ferrier's first experiments on the sensory cortical centres, but inasmuch as the second edition of his great work on "The Functions of the Brain" was not published until after these papers had appeared in the *Medical Chronicle*, I shall content myself with saying that it contains a profoundly interesting article (p. 441, *et seq.*) on the subject of aphasia generally.

which he had recently suffered. The language in which this information was imparted may be gathered from the following note which he
handed to me, and which was evidently a carefully prepared account of
his seizure and subsequent symptoms :—

> Done 10 years Dr. Taylor Guy's Hospital London, but same night back to Man-
> chester. Last month Guy's Hospital 3 weeks Dr. Goodhart and no better and now
> near done, only my Health little longer, and see my Lungs and Throat and four
> weeks bad Pluracy. But try no more tonge.

The vocal speech of this patient was more disconnected, if possible,
than his written speech. During my examination of him nothing could
restrain his volubility; he poured forth volley after volley of words,
many of which I was quite unable to understand, but aided by my
knowledge of the symptoms which usually usher in and accompany
aphasic disorders, and by his expressive gestures, I have no doubt I
formed a tolerably correct idea of the information he intended to impart.
This patient, with his incessant and disjointed talk, reminded me
forcibly of an active and restless child who from his continuous prattle
earns the name of "chatterbox," and indeed a careful examination shows
that the vocal speech of the former is by no means unlike that of the
latter. In both instances the articles, conjunctions, prepositions, and
auxiliary verbs are omitted, while nouns are generally used in preference to
personal pronouns. The vocal speech of this patient is also very similar in
its construction to languages, like the Chinese, which have only reached
the *radical* stage of development. The Chinese[1] language has neither
declensions nor conjugations, and it depends upon the position which a
word occupies in a sentence whether it has the signification of a verb, a
substantive, an adjective, an adverb, or a preposition. It is clear that
in a language of this kind the meaning which the speaker intends to
convey must be interpreted largely by means of his intonations and
gestures. It appears to me, therefore, that the vocal speech of the
patient whose case has just been mentioned, manifests a dissolution of
language to the level of the forms which preceded the *inflexional* stage
in the development of all languages, or to the level of the vocal speech
of early childhood.[2] And, indeed, the post-mortem records of cases of

---

[1] See Müller (Max), "Lectures on the Science of Language," 2nd Series, Lecture II., p. 84 *et
seq.*; and Kussmaul (A.), Art. "Disturbances of Speech."—*Ziemssen's Cyclopædia*, Vol. XIV.,
1878, p. 791.

[2] The relation subsisting between the Chinese language and that of childhood is thus
clearly stated by Prof. Max Müller: "If we watch," he says, "the language of a child, which is
in reality Chinese spoken in English, we see that there is a form of thought and of language
perfectly intelligible to those who have studied it, in which, nevertheless, the distinction between
noun and verb—nay, between subject and predicate—is not yet realised. If a child says *up*, that *up*
is, to his mind, noun, verb, adjective, all in one. It means, 'I went to get up on my mother's
lap.' If an English child says *ta*, that *ta* is both a noun, thanks, and a verb, I thank you. Nay
even if a child learns to speak grammatically it does not yet think grammatically; it seems in

what Dr. Broadbent calls "loss of nouns," and "loss of the power of expressing the relation between things," shows that the lesion occupies nearly the same position in the brain in both cases, proving that there cannot be any essential difference between the two affections. In the case of "loss of nouns," reported by Dr. Broadbent,[1] the two posterior of the six convolutions of the Island of Reil "had entirely disappeared, leaving a smooth surface of a pale fawn colour, formed apparently by the external capsule, itself somewhat degenerated, but overlying grey substance of the lenticular nucleus, which was not appreciably altered. The angular gyrus round the extremity of the Sylvian fissure, and the supra-marginal lobule forming its upper margin posteriorly, were undermined by an extensive area of degeneration continuous with that which had destroyed the posterior convolutions." Other spots of degeneration were found in the basal ganglia and the substance of the left hemisphere, but these had evidently not taken part in causing the disorder of speech.

A good example of the speech disorder, which Dr. Broadbent names "inability to express the relation between things," and in which a post-mortem examination was obtained, is the case reported by Lichtheim,[2] which we have already mentioned. The patient had suffered from an attack of right-sided hemiplegia and aphasia, from the former of which he made a partial recovery, but the latter persisted. He understood both spoken and written speech, and could repeat short words correctly, but made mistakes in repeating sentences. His writing was very imperfect, but he could copy correctly. "When asked to relate his history," says the report, "he strings together in a fluent manner numerous words, of which scarcely one now and then can be made out. The following were noted : ' Evening, five and twenty, and.' Patient is aware of the incorrectness of his diction, and tries to assist himself with gestures. He succeeds better with single short words and answers ; thus, in answer to the question, ' What was there for supper ?' he answered, ' Bread, meat, potatoes,' with only two mistakes. His own name he mutilates." The following were the chief changes found at the autopsy. " There is considerable depression of the convolutions bordering the upper and posterior part of the left Sylvian fissure, as well as of the ascending frontal and parietal convolutions. The Island of Reil is sunken,

speaking to wear the garments of its parents, though it has not yet grown into them. A child says, 'I am hungry,' without an idea that it is different from '*hungry*,' and that both are united to an auxiliary verb, which auxiliary verb again was a compound of a root *as*, and a personal termination *mi*, giving us the Sanskrit *asmi*, I am. A Chinese child would express exactly the same idea by one word, *shi* to eat, or food. The only difference would be that a Chinese child speaks the language of a child, an English child the language of a man."—*Loc. cit.*, p. 86.

[1] Broadbent (W. H.) "On a particular form of amnesia. Loss of nouns."—*Medico-Chirurgical Transactions*, 2nd Series, Vol. XLIX., 1884, p. 249.

[2] Lichtheim (L.) "On Aphasia." Translated by Dr. A. de Watteville. *Brain*, Vol. VII., 1884-5. p. 445.

forming a depression into which the second frontal convolution falls suddenly to a depth of 1½ cm. ; the third frontal convolution terminates into it likewise. Here the pia is inflamed with yellow discoloration. The depression is bounded posteriorly by the fissure of Sylvius. The middle portion of the first temporal convolution is somewhat sunken opposite the depression of the Insula, in which the consistence of the cerebral matter is soft, with harder patches around. . . . The softened patch occupies the bottom of the Sylvian fissure and extends to 1½ cm. from the posterior part of the inferior frontal convolution, and to the neighbouring portions of the ascending frontal convolution. A fragment of the cortex of the ascending temporal is also wanting, but there is no yellow discoloration here." If we now compare the results of the post-mortem examination in these cases, reported by Drs. Broadbent and Lichtheim, it will at once be apparent that the lesion in the one occupies almost the same position as it does in the other. The only differences in the localisation of the lesion that we can discover, are that the area of the softening extended rather further back, round the posterior ends of the horizontal branch of the Sylvian and the parallel fissures (the angular gyrus) in Dr. Broadbent's than it did in Lichtheim's case, while it extended rather further forwards over the inferior portion of the ascending parietal and frontal convolutions in Lichtheim's than in Dr. Broadbent's case. It is also expressly stated in the report of Dr. Broadbent's case, that the angular gyrus and supra-marginal lobule were undermined by degeneration, while it would appear from the report of Lichtheim's case, that the cortex of the supra-marginal lobule was itself softened, and it is, as we have seen, doubtful if the angular gyrus was implicated. But surely such minor differences as have just been noted, would not warrant us in concluding that in Dr. Broadbent's case the lesion had injured the conducting path between a naming and a pro- positionising centre, while in Lichtheim's case the propositionising centre itself was destroyed. In the case of verbal amnesia, already mentioned as having been reported by Rosenthal, the area of softening was found in the second and third temporo-sphenoidal convolutions, and yet this patient also was unable to name objects at sight, and had, therefore, lost the use of concrete nouns, but in Dr. Broadbent's case the lesion was found in the supra-marginal lobule and angular gyrus. If then the clinical form of aphasia, named "loss of nouns," is caused by rupture of the conducting path which unites a naming centre with a propositionising centre, this path is capable of being injured either by a lesion in the second and third temporo-sphenoidal convolutions, or in the supra-marginal lobule and angular gyrus. Nothing in the structure of the brain, or in the direction of the association fibres, is known to me which could be held to justify such a supposition.

The following appears to me to be a much more reasonable interpretation of the symptoms of these forms of aphasia. The first temporosphenoidal convolution (the auditory sensory centre) is in direct relation with the central end of the fibres of the auditory centripetal conducting paths, and the supra-marginal lobule and second and third temporosphenoidal convolutions are in indirect relation with them. Now the understanding of spoken speech is first developed in healthy persons in connection with the sense of hearing, and consequently partial damage of that portion of the centre which is connected with the auditory conducting paths will give rise to partial disorders of speech, while complete disorganisation of this portion of the centre will give rise to complete loss of the faculty of understanding vocal speech. Now the least damage to the centre will be done by lesions of those portions of the cortex which are indirectly connected with the auditory conducting paths—the second and third temporo-sphenoidal convolutions and the supra-marginal lobule, and the greatest damage to the centre will be done by lesions of that part which is directly connected with this path—the first temporo-sphenoidal convolution. Lesions of the second and third temporo-sphenoidal convolutions will, according to this supposition, only damage the more special part of the great function which has been developed in connection with the sense of hearing, and as the names of individuals, concrete nouns, and highly abstract nouns like "virtue" are the most special part of language, they are the parts of speech which suffer most by such a lesion The case of R. B. (Case 8) is a good example of the kind of speech disorder which is most probably caused by a lesion in these convolutions. The supra-marginal lobule may be supposed to be more intimately connected with the auditory conducting paths than the second and third temporo-sphenoidal convolutions, but the connection between them is still indirect and not direct. Lesion in this locality will therefore leave the fundamental structure of language intact. A lesion of a certain intensity may be supposed to cause loss of all highly abstract and concrete nouns and individual names, and to render the patient incapable of using correctly the moods and tenses of verbs, the inflexions of nouns, and the articles and connecting particles which give definition and precision to language. The patients, however, still retain, as in Lichtheim's case, the use of a few familiar concrete nouns, the names of the simpler qualities and ordinary numerals, a considerable number of verbs, and some other parts of speech which enable them to express their meaning fairly well, especially when aided by gesture. A more profound lesion of the supra-marginal convolution may be supposed to cause loss of all abstract and concrete nouns, like the case reported by Dr. Broadbent. It ought, however, not to be forgotten that a lesion of the supra-marginal convolution often encroaches upon the motor area of the cortex, as in Lichtheim's case, and it is, therefore, probable that in the cases which Dr. Broadbent names "inability to express the relations between things," the sensory disorder of speech is accompanied by some injury of the motor mechanism, a certain justification being thus afforded to Wernicke for giving to such cases the name of *commissural aphasia*. But although we believe that cases of this kind form a sort of transition between the sensory and motor disorders of speech, we find no warrant for the opinion that they are caused by isolated disease of any particular bundle of the association system of fibres, even in spite of the fact that the *fasciculus uncinatus* which passes across the bottom of the Sylvian fissure to connect the temporo-sphenoidal with the frontal lobe is likely to be involved in the lesion. There is no case known to me in which disease of this bundle of fibres caused a disorder of speech when the cortex was intact.

The first temporo-sphenoidal convolution has so far been supposed to be intact, and although the patients had lost the power of naming things they still understood the noun when uttered in their hearing. When, however, this convolution is itself diseased word-deafness is superadded to the other disorders of language. If the convolution is only partially destroyed the patient can understand much of what is uttered in his hearing, so long as the sense of the statements does not depend upon concrete nouns, and he is also able to express many of his own thoughts by means of pronouns, simple verbs, and adverbs of time and place, aided by gestures. The case of James Lee[1] (Case 5), is a good example of this speech disorder. If the first temporo-sphenoidal convolution is thoroughly disorganised, the patient fails to understand almost all words which are uttered in his hearing, and he is likewise unable to express his thoughts in articulate speech, although he is still able to repeat "recurring utterances," like the

[1] Such patients have also the use of a few general and abstract nouns. James Lee (Case 5), whose speech disorder is a very good example of the one under consideration, made frequent use of *matter* and *gentlemen*, but those words were generally used when he was endeavouring to describe his own bodily feelings and previous experiences, and seldom in reference to any one thing or person presented to him. The words were never, indeed, used singly, but always as part of a sentence. It seems clear, therefore, that the patient is aided in his utterance of such words by their association with others, and it is probable that it is this association which revives them in memory, and not any external impression. The power of association in suggesting complex words to persons suffering from sensory aphasia is well illustrated by the capacity James Lee manifested of counting from one up to a hundred, although unable to utter one word of the series unless he started with the first. Dr. Gairdner believes that in associations of this kind the words are set free "very much as one wholly unskilled in music might let loose the tunes of a *barrel-organ* without any direct cognisance of the music he is playing, or the precise combinations of movement he is calling into action." But whether this be the explanation or not, there can be no doubt that the patient is, as Dr. Gairdner says, capable of uttering phrases *en masse* which he is incompetent to utter in detail.—*Loc. cit.*, p. 19, and *British Medical Journal*, Vol. II., 1885, p. 310.

phrases "John, if you please," "thank you, ma'am," used by John Morris (Case 6). When the first temporo-sphenoidal convolution is completely disorganised, the patient is quite unable to understand any word uttered in his hearing, and even the power of repeating recurring utterances fail him, but he is still capable of uttering a larger number of vocalisations to which no definite meaning can be attached, this speech disorder constituting the variety which we have already named *syllabic paraphasia*. When, however, gibberish aphasia is caused by a lesion in the sensory area of the cortex, without any damage being at the same time done to the motor mechanism of speech, it is probable that disease of the angular gyrus must necessarily be associated with that of the first temporo-sphenoidal convolution. At any rate, in a case reported by Dr. Broadbent,[1] in which the patient talked incomprehensible gibberish, he was unable to comprehend either written or spoken language, and at the autopsy an extensive area of softening was found in the angular gyrus, supra-marginal lobule, and first temporo-sphenoidal convolution.

In the conditions already described under the names of partial perceptive blindness, and partial perceptive deafness, it is highly probable that the angular gyrus and the first temporo-sphenoidal convolutions respectively are diseased in both hemispheres. These affections would, therefore, bear a similar relation to the other forms of sensory aphasia, that the bulbar paralysis which results from lesion of the posterior part of the third frontal convolution on both sides bears to the other forms of motor aphasia. A disorder of the apperceptive faculty of speech is, so far as we know, not caused by disease of the sensory conducting paths, in correspondence with the motor aphasia and bulbar paralysis, which result from injury of the motor conducting paths in their passage through the internal capsules, crura cerebri and pons. Hemianæsthesia is, however, not an unfrequent accompaniment of sensory aphasia, but this symptom is certainly caused in some cases, and probably in all, by a second lesion in the optic thalamus or lenticular nucleus, whereby the sensory conducting path is injured in its ascent to the cortex in the posterior part of the internal capsule.

The theory which has just been advanced still encounters one serious difficulty. In the second degree of verbal amnesia, the patient, as in the case of R. B. (Case 8), is unable to name most objects presented to him, but he immediately identifies the object when the name is uttered in his hearing, or presented to him in writing. Why then should the object fail to revive the idea of the verbal sign, while the verbal sign revives the idea of the object? The reply to this question

[1] Broadbent (W. H.) "On a case of amnesia, with post-mortem examination."—*Medico-Chirurgical Transactions*, Vol. LX., 1878, p. 147.

is to be found in an analysis, not of language, which is the *product* as well as the instrument of thought, but in an examination of the *process* of thought itself. Our thoughts may be said roughly to move on two extreme, and a middle plane, with every intermediate level. On the low level our thoughts correspond more or less to the thoughts which may be supposed to pass through the minds of the lower animals. We shall not waste time in proving that animals think. A dog, for example, on seeing his master from a distance, infers that if he will run some paces he will experience certain sensations of resistance, of touch, and of smell, or in other words, the presented sensation caused by the impression of certain rays of light falling on the dog's eye has revived a large group of represented sensations which have always been associated with the presented one in his experience. This shows that the dog has formed a perception of his master, just as a fellow man would have done had the same rays of light fallen upon his eye. The perceptions which a dog and a human being looking at a man from the same distance form of him will differ doubtless greatly in detail, but the fact that the respective perceptions prove adequate for the practical guidance of both animals under similar circumstances shows that they must be fundamentally alike. A dog not only forms perceptions of objects, but there can be little doubt that ideal revivications of perceptions previously experienced pass through his mind. Dogs not unfrequently bark excitedly in their sleep, and it is, therefore, highly probable that they are dreaming of being engaged in the pleasures of the chase, or in resisting an attack from an enemy. And if dogs dream it may be inferred that reminiscences of previously experienced pleasures and dangers, with all their attendant circumstances, pass through their minds in the waking state. It may, therefore, be concluded that animals think, not only in trains of *feelings*, as Professor Huxley asserts, but in trains of *perceptions*. Now a great part of our thinking consists of perceptions experienced at the moment, and of reminiscences of previously experienced perceptions. Suppose one takes a walk in a green field, he perceives the greensward, the trees, and the foliage, from which rays of light are reflected to his eyes, and the cattle and birds, whose lowing and singing fall upon his ear, and the more vivid of these perceptions start a train of reminiscent perceptions, probably from scenes in his childhood, while there is with him an ever-present feeling of the great conditions of all our sensuous perceptions—time and space. It is possible that in addition to these thoughts he may carry on a train of reasoning in words, but whether that be so or not, there can be no question that the greater part of his thinking has consisted of perceptions actually experienced, and trains of perceptions ideally revived. Thinking by means of *percepts* is, therefore, common to man with the lower animals, and although the thoughts of the former, even on this level, are much more varied and complex than those of the latter, yet there is no essential difference in principle between the two.

Let us now pass at once to the consideration of the other extreme plane of thought. Thinking, on the high level, has for its subject-matter the relations in which phenomena are presented to us when they are detached from all individual perceptions, and it can only be carried on by means of mathematical symbols and *abstract* nouns. It is quite manifest that thinking of this kind is beyond the reach, not only of the lower animals, but also of the lower races, and the youth and uneducated of the higher races of mankind. Such words as "function" and "co-efficient" in mathematics, and "virtue" and "benevolence" in morals, are not even capable of being expressed in the languages of the inferior races, and can only have a vague meaning to all except those whose mental faculties are fully developed and highly educated. Thinking on the middle level is possible to every healthy man who has passed the age of early infancy, but not to any animal. It is carried on by means of words, and chiefly by the aid of general names, or *concepts,* and it is scarcely necessary to add that thinking of this kind admits of innumerable degrees. In the early stages of development the concept and the precept must have almost been identical. Aboriginal man, even before he had any general name for his fellow man, was, no doubt, able to form a perception of him as the lower animals do. On hearing a footstep, for example, the image of one of his fellow men, or of several of them in succession, would arise in his mind, and the rapidity with which images of different people passed through his mind gave to the perception an element of generality. But let us suppose that an aboriginal man had advanced so far in the acquirement or construction of language as to have a word *man,* as in the Sanskrit, signifying *to think,* and that he had now applied it to designate the animal who is pre-eminently endowed with the power of thinking, the word, on being uttered by himself or another, would not only call up an image of a fellow man, as the footstep did, but by the aid of it the power of generalising would be indefinitely increased. The verbal symbol would now stand for generations of men yet unborn, and for those that had already died just as well as for the men that were living around him. Once the *concept* man had become detached from the *percept* man the meaning of the former has gone on widening in *intent,* if not in *extent,* along with the growing knowledge of our own minds and of the universe by which we are surrounded, while the signification of the latter has remained comparatively, although not altogether, stationary. In order to prove this we have only to glance at the meaning which an educated biologist would attach to the general name *man.* According to Prof. Huxley,[1] man is one of the orders of mammalia having a discoidal deciduate placenta, the hallux provided with a flat nail, a particular dental formula, and certain other peculiarities. It will at once be apparent that this definition can have no meaning to anyone except he is possessed of a wide knowledge of the structure of other animals, and this presupposes a knowledge of the properties of the substances which enter into the composition of the bodies of animals; in short, the definition can have no meaning except as a part of a complete scheme of knowledge of man and of the universe in which he lives. And although only a very few men can form such a conception of man as Prof. Huxley, yet the conception which each man forms of his kind is determined by his particular scheme of knowledge of the universe at large. Concepts as embodied in general names are generalisations of innumerable percepts of various orders, and as such they are altogether on a higher plane of thought than the individual percepts aroused in our minds by the exercise of our senses. Man lived upon this planet thousands upon thousands of years before it was possible for him to regard himself in the light of Prof. Huxley's definition, and the definition itself was only rendered possible by the accumulated knowledge of ages. It will thus be seen that the passage from the *percept* to the *concept* has been an exceedingly slow and gradual process; but each step in the growth of the latter presupposes that it can be readily applied for the practical purpose of recognising the former when presented to the senses. In other words, the passage from the low level of thinking in *percepts* to the middle level of thinking in *concepts* has been a slow and difficult one, but each acquirement on the middle level presupposes a ready and easy transition to the low level.

Let us now apply this somewhat crude theory of the process of thinking to the elucidation of the phenomena of sensory aphasia. We may at once state our disbelief in the possibility of parcelling out the nervous structure in the cortex of the brain, the activity of which is the correlate of the process of thought, into distinct centres, each having a separate faculty, such as naming, ideation, and propositionising. There is nothing in the structure of the brain to warrant such a supposition. The cortex is spread out like a mantle over the surface of the hemispheres, and to this structure fibres ascend from the surface of the body and terminate in or amongst the superficial cells, while other fibres emerge from the inner layer of the cortical cells and pass out to end in the muscles. The centripetal fibres are gathered into groups, which terminate chiefly but not entirely in the posterior and inferior portion of the cortex of each hemisphere; and the centrifugal fibres, also collected

[1] Huxley (Thomas Henry). "An introduction to the classification of animals." Lond., 1869, p. 99.

into somewhat indefinite groups, emerge from the middle and anterior portions of the cortex, and consequently we have no objection to call the portions of the cortex in which the former groups terminate, and from which the latter emerge, sensory and motor centres respectively. In addition to the ingoing and outgoing fibres, the cortex is furnished by a system of association fibres which connect different parts of the cortex with one another.

The annexed diagram, sketched for me with his usual kindness by my friend, Prof. A. H. Young, will help the reader to understand this view of the cortex. From the eye and ear centripetal fibres (v and a) ascend to

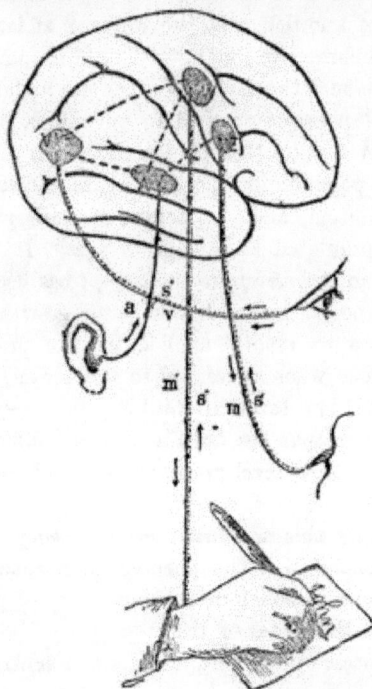

terminate in the angular gyrus (V) and first temporo-sphenoidal convolutions (A) respectively, but in reality these fibres are directly connected with a much larger area of the cortex than is here indicated. In addition to these, fibres of muscular sense (s s' and s''), indicated by dotted lines, ascend from the muscles of articulation, from those of the hand, and from those of the eyeball to reach the cortex. These fibres are represented as terminating on the surface of the motor centres of the cortex, but it is doubtful whether they ought not be made to terminate along with the cutaneous sensory nerves—not represented in the figure—in the hippocampal region. The dotted lines which connect the sensory centres with one another and with the motor centres represent the association system of fibres, but they have no pretensions to anything like accuracy of representation. The centres of vocal and written expression are represented at E and W, and these are connected by means of centrifugal fibres m and m' with the vocal apparatus and hand respectively. According to this scheme it is possible to speak of motor and sensory centres, but at most there can be only one perceptive centre. A moment's consideration will make this plain to any one. Suppose that the rays of light which are reflected from the surface of an orange fall upon the eye. As long as these rays arouse in the mind a sensation of colour alone, so long is the cortical excitation restricted to the visual centre. But if once a perception of the cause of the colour is formed, then the object is thought of as having a peculiar smell and taste, as being solid and having weight, and as being a certain distance from the eye, which shows that the excitation has extended from the visual centre to the sensory centres of smell and taste, and of the muscular sense of the hand and of the eyeball respectively. The act of perception, therefore, is correlated with the active excitation of one sensory centre— say the visual—and faint excitation of several other sensory centres, and consequently the perceptive centre must be formed by the union of all the sensory centres. The simple relations between things are perceived in the same way, and they also are correlatives of the activity of the one perceptive centre. If the eye be directed from the orange on the table to a book, the new excitation of the perceptive centre gives to the mind the shock of *difference*, and if it be directed from the one orange to another the sensation of *sameness* is experienced, although these relations are not thought of apart from the perception of particular objects, until the middle and high levels of thinking are attained.

And on passing from thinking by percepts to thinking by concepts, and from that to thinking by abstracts, there are no new centres introduced, but only a complication upon complication of the one perceptive centre. All that can be said is that the correlative of perceptive thinking is excitation of that portion of the cortex of the brain which is directly connected with the sensory inlets, of conceptive thinking excitation of portions of the cortex which are indirectly connected with them, and of abstract thinking excitation of portions which are still more remotely connected with them. It must, however, be remembered that the effective working of the portions of the cortex which are remotely connected with the sensory inlets will, in a great measure, depend upon the integrity of those which are in direct relation with them.

Let us now attend to the effects of dissolution of this structure. A destructive lesion of the portions of the cortex which are most remotely connected with the sensory inlets would destroy the capacity of the patient for highly abstract reasoning, and would no doubt inflict considerable damage on the language in which abstract thought is embodied, but this condition would not be recognised as an aphasia; and even the intermediate portions of the cortex in which conceptive thought is carried on might be seriously damaged without giving rise to a special speech disorder, inasmuch as any impairment of speech which might be present would only be regarded as a part of a general decay of the reasoning faculties. When, however, the lesion is situated in or near to the sensory inlets, a disorder of language results which is out of all proportion to the general impairment of the reasoning faculties. There are several reasons why the power of thought is comparatively spared under such circumstances. Thinking by means of percepts is correlated with the activity of both hemispheres, and consequently one hemisphere will carry on thought on the low level when the other is injured. And although speech is organised in one hemisphere, a destructive lesion of one of its sensory inlets does not cut off the patient altogether from communication with the external world. The portions of the cortex the activity of which is correlated with thinking by concepts and abstracts can be reached in the word-deaf through the eye, and in the word-blind through the ear, and in those who are both word-deaf and word-blind through the nerves of muscular sense and those of the other special senses, so that thinking on the middle and high levels is not completely arrested. Inasmuch as speech is first organised in connection with the sense of hearing, a lesion in or near the auditory cortical centre will cause, as already remarked, a greater disorder of speech than does disease of the sensory centres. Now, suppose that the auditory centre itself is spared, and the lesion is situated in the cortex near to it. The patient can now appreciate a general name uttered in his hearing as an acoustic image, and he can immediately repeat it. The verbal sign has, however, never sounded in his ear without reviving some kind of concept, the two being inseparable. The concept evoked by the verbal sign may be, indeed, and probably is, a very undeveloped and imperfect one, but it suffices for the purposes of identifying the corresponding percept. When, however, the excitation of what remains of the auditory centre, which is caused by the falling of the verbal sign on the ear, fades away, the percept is not capable of reviving in memory either the concept or the general name which embodies it. We see nothing in this except that less resistance is offered by the nervous structures to the passage of nerve currents from the less organised structure representing the concept to the more organised one representing the percept, than from the more organised to the less organised, but this is only a particular example of a general law. There is no occasion, therefore, to postulate the existence of a complicated mechanism of centres and conducting paths for what can be so simply accounted for.

The diagnosis of the nature of the lesions which give rise to aphasia is most profitably considered in treatises on nervous diseases or on general medicine, and the subject will consequently be passed over here. The prognosis in different cases, and the general treatment of the patient, depending as they both do on the nature of the lesion, will also not be discussed in these pages. In bringing the work to a close we shall content ourselves with making a few remarks on the special treatment of aphasia considered as a symptom and without reference to the lesion which has given rise to it. In pure motor aphasia the patient understands what is said in his hearing, or presented to him in written or printed characters, but he is unable to express himself in articulate speech or in writing. So far as writing is concerned the left hand is capable of being educated by a series of exercises, just as the right was educated in childhood. With articulate speech the process of education must be conducted on the same principles as in writing; that is, by imitating the method adopted by the mother, often more or less unconsciously, in training her infant to speak. The first lesson consists in getting the patient to utter almost any kind of vocalisation, and the hearing of the sound of his own voice gives him so much pleasure and encouragement that he soon learns to utter and recognise many of the vowel sounds. The patient should then be led to imitate the movements of the lips of his instructor, and no great difficulty will be experienced in teaching him how to utter the explosive consonants first, and other consonants subsequently.[1] After some of the vowel sounds and consonants are mastered the patient can then be taught to articulate short monosyllabic words; and in selecting the words to be taught to the patient care should be taken to use pronouns and monosyllabic verbs, these being the most useful for the construction of short phrases.

In sensory aphasia, the kind of training which will be found most suitable will depend upon whether the visual or auditory functions are damaged. In word-blindness the patient may learn to read by the aid of tactile and muscular sensations. A case of word-blindness is recorded by Westphal,[2] in which the patient re-acquired the power of reading his own writing by passing his finger over each letter as if he were re-writing them, and Magnan taught a patient to read by the device of raised letters.[3] The case of a gentleman, recorded by Charcot, has already been mentioned in these pages,[4] who, after an attack of word-blindness,

[1] See Bristowe (John Syer). "The Physiological and Pathological Relations of the Voice and Speech." London, 1880, p. 114, et seq.

[2] See Kussmaul (A.) *Ziemssen's Cyclopædia*, Vol. XIV., 1878, p. 770.

[3] See Skwortzoff (Nadan). "De la Cecité et de la Surdité des Mots dans l'Aphasie." Paris, 1881, Obs. VII., p. 84. Also Ferrier (David). "The Functions of the Brain," Second Edition, 1881 p. 456.   [4] Page 75.

taught himself to decipher written or printed words with considerable rapidity by moving the index finger of his right hand in the air as if he were writing.

In word-deafness the education of a new faculty is carried on with considerable difficulty, but such patients soon become very proficient in interpreting gesture language, and systematic education would doubtless greatly increase the range of this power. James Lee (Case 5) was word-deaf, but the power of naming familiar objects, at first altogether lost, had to some slight degree been re-organised in him, and his education might have been carried much further by systematic training. He was so quick in recognising the names of objects when the word was presented to him in writing that had his clothing, and the ordinary objects by which he was surrounded, such as articles of furniture, been distinctly labelled, he would very soon have re-acquired the names of these objects. The patient had jotted down for himself in a pocket-book the names which were most urgently required by him, and a carefully selected list of concrete nouns, alphabetically arranged, would doubtless have been of much service to him. The service which such a list may render to the word-deaf is well illustrated in a man whose case was reported by Graves.[1] This man had lost his memory for proper names and substantives in general, with the exception of the first letter of the noun; but he prepared for himself an alphabetically arranged dictionary of the substantives required in his home intercourse, and whenever it became necessary for him to use a noun he looked it up in his dictionary. When he wished to say "cow" he looked under C, and as long as he kept his eye upon the written name he could pronounce it, but was unable to do so a moment afterwards. While examining James Lee (Case 5) one day an incident occurred which shows that he also was sometimes aided in recalling a name by remembering its first letter. He was asked to name his hand, and, as usual, he looked in the distance as if he were listening intently, but after a time he applied the tip of the index finger of the right hand to the palm of the left, and moved it as if he were writing the capital letter H in a current hand, and he immediately called out "Hand! that is my hand, the whole of it." "The whole of it" was doubtless meant to imply that the name applied to the hand in contra-distinction to the fingers, which he was frequently asked to name separately. In the combined forms of sensory and motor aphasia the treatment becomes difficult and uncertain, but the education of the impressive must necessarily precede that of the expressive faculty.

[1] Graves. *Dublin Quarterly Journal*, Vol. XI., 1851, p. 1, quoted by Kussmaul, *Zeimssen's Cyclopædia*, Vol. XIV., 1878, p. 758.

JOHN HEYWOOD, Excelsior Steam Printing and Bookbinding Works, Hulme Hall Road, Manchester.

www.ingramcontent.com/pod-product-compliance
Lightning Source LLC
Chambersburg PA
CBHW030607270326
41927CB00007B/1086